HUMAN SEXUALITY

HUMAN SEXUALITY
Sense and Nonsense

Herant Katchadourian
STANFORD UNIVERSITY

W. H. FREEMAN AND COMPANY
San Francisco

Library of Congress Cataloging in Publication Data

Katchadourian, Herant A.
Human sexuality.

Reprint of the 1972 ed. published by Stanford Alumni Association, Stanford, Calif., in series: The Portable Stanford.
Includes bibliographies.
1. Sex (Psychology) I. Title. [DNLM: 1. Sex manuals. HQ31 K19h 1972a]
BF692.K33 1974 155.3 74-8600
ISBN 0-7167-0584-2
ISBN 0-7167-0583-4 (pbk)

Printed in the United States of America

1 2 3 4 5 6 7 8 9

This book was published
originally as a part of
The Portable Stanford,
a series of books published by
the Stanford Alumni Association.

CONTENTS

ILLUSTRATIONS

PAGES 3, 6, 9, 12 — Figures 1 through 6 and figure 8 are redrawn, with permission, from Dienhart, *Basic Human Anatomy and Physiology*, Second Edition. Philadelphia: W. B. Saunders Company, 1973.

PAGES 12, 19 — Figures 7 and 9 are from Crawley, Malfetti, Stewart, and Dias, *Reproduction, Sex, and Preparation for Marriage*. Copyright 1964. Reprinted by permission of Prentice-Hall, Inc., Englewood Cliffs, New Jersey.

PAGE 24 — Figures 10 and 11 are reprinted by permission from Masters and Johnson, *Human Sexual Response*. Boston: Little, Brown and Company, 1966.

PREFACE

Sex itself hardly needs justification, but books about it certainly do. Given an estimated 1,500 different "sex manuals" already on the market, one is inclined to throw up one's hands and exclaim, "*Another* book on sex?" Not only do I share such consternation, but having recently coauthored one 500-page textbook on sex, at first I was about as eager to write another book on sex—this one—as I was to climb the Sierra in my pajamas. Why bother? I am told that literary floods usually indicate one of two things: either there is a great deal to be said about the subject, or, the facts being anybody's guess, lack of pen and pad is all that separates the enterprising from authorship. My persuaders argued that inasmuch as I had just gone through the literature in preparation for *Fundamentals of Human Sexuality*, the text Donald Lunde and I wrote, impressions would be fresh in my mind. Why not make these available to readers who require a book less comprehensive than *Fundamentals of Human Sexuality*? So be it.

In keeping with the aims of the Stanford Alumni Association, which published this book originally as a volume in its Portable Stanford series, *Human Sexuality* was designed to fulfill two roles: first, to acquaint the reader with the salient facts in the area of human sexuality; second, to provide guidance in the often frustrating search for the "right books." Although this brief book is not intended to be a comprehensive and in-depth coverage of the field or any part of it, I will try to say something about most of the major facets of sex. I will assume that some readers will not go beyond this book (including those who do not even finish it!), and that others will pursue the matter further to varying lengths.

Stanford, California
April 1974

Herant Katchadourian

INTRODUCTION TO THE LITERATURE ON HUMAN SEXUALITY

A few words about the literature on human sexuality and what portion of it will be covered here: I shall occasionally refer to, but will make no attempt to outline, literary and popular contributions to this area. The literary works are too voluminous, and the popular approach is hardly worth outlining here. Therefore, I shall concentrate on the specialized and scientific (or what one *hopes* are scientific) writings on the subject.

The literature on the physical basis of sex is found primarily in the field of medicine and its allied sciences. The anatomy of the sexual organs, for instance, is treated like the anatomy of other body systems, and knowledge about it has accumulated progressively through the long history of medicine. The study of sex hormones, like endocrinology itself, is a more recent rapidly expanding field.

In the investigation of the physiology of sexual functions, medicine has sadly lagged behind. If you examine standard physiology texts that deal in excruciating detail with various functions of the body, you will find only a few meager pages on fundamental sexual functions like orgasm. Some of the elementary facts about orgasm were unknown until the recent investigations by Masters and Johnson, the first concentrated research in this area. The reason for this selective neglect is quite clear. The study of sexual functions entails the study of individuals experiencing sex. You can study anatomy by dissecting a cadaver, but to study orgasm you need a live body—and that raises complex social problems about the propriety of such research.

For this reason, the study of sexual behavior is still in its infancy. The roots of sex research go back to the post-Enlightenment era when it gradually became legitimate to study empirically what people actually do, rather than to speculate endlessly on what they ought to be doing. Writers of the nineteenth century nevertheless tended to dwell upon abstractions like the "nature of sexuality." There was also a heavy reliance on material from classical periods and ethnological accounts, since it was safer to talk about what people did centuries ago or thousands of miles away rather than about what people were doing there and then in the writer's own back yard.

It was inevitable, however, that matters closer to home would ultimately come under scrutiny. One of the eminent pioneers in sex research was the Victorian English physician Henry Havelock Ellis (1859-1939—not to be confused with Albert Ellis, a contemporary American psychologist and prolific writer in this field). Using a narrative approach, Ellis compiled everything he could find on sexual behavior from unpublished case histories and correspondence as well as previously published material. The result, a seven-volume compendium, was written, revised, and expanded from 1896 to 1928. This work has now been reissued, unabridged but in a two-volume format. For the historically minded reader, it contains a wealth of fascinating information (*Studies in the Psychology of Sex*. New York: Random House, 1942).

Despite his medical training, Ellis was not afflicted with the usual preoccupation with pathology—which cannot be said for his contemporary, Richard von Krafft-Ebing (1840-1902). Krafft-Ebing also shared the passion for classification typical of the German medicine of that period. The result was *Psychopathia Sexualis* (1886), the first serious attempt at systematizing sexual pathology. Although this book is still in circulation, it is seriously outdated and should be considered a work of historical interest only.

A third man in this group whose life and work spanned the transitional period from the past to the present century was Sigmund Freud, who was born three years before Ellis and died the same year as Ellis. Unlike Ellis and Krafft-Ebing, Freud was not primarily a sex researcher in that he did not deliberately set out to study sex but was led to it through his clinical work with psychiatric patients. Nevertheless, sexuality in a vastly expanded sense was to become the focal point in his theories of psychosexual development and mental functioning and to exert a vast influence upon our thinking in innumerable areas. Freud's collected works, which run to more than twenty volumes, are imbued with sexual themes.

One other man whose name is a household word in connection with sex research is Alfred C. Kinsey. Kinsey, a modern scientist born in 1894, spent the first two decades of his professional life as an entomologist. His interest in sex research began when he was over forty years old and was motivated by his inability to find adequate answers to the questions on sex that his students asked him. Starting tentatively by inquiring about what people actually did in their sex lives, Kinsey and his associates ultimately gathered more than 16,000 case histories. Kinsey himself collected 7,000 histories, which amounts to fourteen histories a week for ten years. His untimely death interrupted his plan to expand his sample to 100,000 individuals. Kinsey's data are flawed; his sample was not random or sufficiently representative, and questions have been raised about the reliability of interviewing as an information gathering method in this field. Nevertheless, his data are still the most comprehensive figures available, and as long as they are used responsibly they continue to be a rich source of valuable knowledge.

For additional information about the contribution of the above figures and the history of sex research, see "Advances in Modern Sex Research" by Leo P. Chall in *The Encyclopedia of Sexual Behavior* edited by Albert Ellis and Albert Abarbanel (New York: Hawthorn Books, Inc., 1967). A more detailed but rather opinionated work is *The Sex Researchers* by E.M. Brecher (Boston: Little, Brown and Company, 1969). This work has comprehensive biographical references to Ellis and Freud, but there is no biography of Krafft-Ebing. The first detailed biographies of Kinsey have recently been published (C.V. Christenson, *Kinsey, A Biography*. Indiana University Press, 1971, and Wardell B. Pomeroy, *Dr. Kinsey and the Institute of Sex Research*. New York: Harper and Row, 1972).

In connection with the "reader's guide" function this booklet is designed to fulfill, I will provide three types of references. First, at the conclusion of this introduction, I will discuss a few books that deal with human sexuality in a comprehensive manner. For the reader with general interests, these will be the sources to read next. Second, at the back of the book, in the Reader's Guide section, I will list references pertaining to each topic area. Third, within the context of various discussions, references will be used to support some particular issue or point.

Although for all the major aspects of sex I will mention several references in increasing order of complexity, I will try to avoid redundancy by not listing works that accomplish the same task. Some books that are left out are possibly of comparable merit to those mentioned; thus, while I do recommend by inclusion, I do not intend to depreciate by exclusion. So far as possible, these judgments will be based on my understanding of how a book is viewed by people knowledgeable in the field. When the choices reflect my idiosyncratic references, I will remember to say so.

At the risk of appearing to be blatantly self-promoting, I am going to recommend *Fundamentals of Human Sexuality* (H.A. Katchadourian and D.T. Lunde. New York: Holt, Rinehart, and Winston, Inc., 1972). Faced with the lack of adequate textbooks in this field, fellow Stanford psychiatrist Donald T. Lunde and I prepared this text for college-level courses such as the one we teach at Stanford. Its eighteen chapters cover the major biological, behavioral, and cultural aspects of human sexuality. There are guest chapters on the erotic in literature and film by Strother B. Purdy and on the erotic in art by Lorenz Eitner. This embarrassing exercise of pushing our own book is made somewhat easier by my belief that there is currently no serious competitor to it and also by my guess that, were you presently a Stanford undergraduate, there is a good chance you would be taking the course in human sexuality and reading our book anyway!

A useful reference work previously mentioned here is *The Encyclopedia of Sexual Behavior*. Although it is not exactly encyclopedic in scope, this book includes 112 chapters by 98 authors representing some of the best-known names in this field. The choice of topics and quality of coverage are uneven, but there is enough that is worthwhile to justify its endorsement as a general reference.

The two Kinsey volumes are research monographs, and I hesitate to include them here. But because they are classics in this field and are packed with vast amounts of information, you may consider reading them even though you will probably bypass the masses of tables and statistics. (A.C. Kinsey, W.B. Pomeroy, and C.E. Martin. *Sexual Behavior in the Male*. Philadelphia: W.B. Saunders Company, 1948. A.C. Kinsey, W.B. Pomeroy, C.E. Martin, and P.H. Gebhard. *Sexual Behavior in the Female*. Philadelphia: W.B. Saunders Company, 1953.)

Not exactly a general reference, but a most worthwhile book, is *Patterns of Sexual Behavior* by C.S. Ford and F.A. Beach. (New York: Harper and Row, 1951.) Although much of the data in this book are outdated, the vast panorama of cross-cultural and cross-species patterns of behavior it surveys does not suffer too badly as a result. If a particular cultural pattern has changed, one can still learn from the past.

Finally, you may wish to consider several readings of general interest that have recently appeared in paperback: B. Lieberman, *Human Sexual Behavior: A Book of Readings*. New York: John Wiley and Sons, Inc., 1971; J.L. Malfetti and E.M. Eidlitz, *Perspectives on Sexuality: A Literary Collection*. New York: Holt, Rinehart and Winston, Inc., 1972; R.R. Bell and M. Garden, *The Social Dimension of Human Sexuality*. Boston: Little, Brown and Company, 1972; J.N. Edwards, *Sex and Society*. Chicago: Markham Publishing Company, 1972; A.McC. Juhasz, *Sexual Development and Behavior: Selected Readings*. Homewood, Illinois: The Dorsey Press, 1973.

HUMAN SEXUALITY

CHAPTER ONE

THE PHYSICAL BASIS OF SEX

BECAUSE THE structure and functions of the organs of the body are inextricably interrelated, one cannot be properly understood without the other. The enormous complexity of living organisms, however, has made it necessary that their study be separated to some extent (and somewhat arbitrarily at times) into various disciplines such as anatomy, physiology, or endocrinology. Although we will follow these divisions in our discussions here, please bear in mind the artificiality of such distinctions: the human body, like all other living organisms, exists and functions as a unitary whole.

Anatomy of Sex Organs

The relationship of anatomy to physiology has been compared to the relationship of geography to history. Like geography, anatomy is the description of the theater where actions unfold. In this regard, two generalizations are in order. First, the sexual or reproductive systems of the male and female are built on the same fundamental ground plan —obvious and cherished differences notwithstanding. Second, in both sexes the genital organs fulfill a double function: the production of germ cells (sperm or eggs) and the production of sex hormones. We shall begin with those functions related to germ cell production and delay discussion of endocrine functions.

Understanding of the sexual apparatus is facilitated if one thinks of the reproductive system as having three structural components: the gonads or organs for the production of germ cells (the testes producing sperm and the ovaries, eggs); a set of tubes for the transport of germ cells and, in the female, for housing the product of their union; and organs for the delivery and reception of sperm (the penis and the vagina).

Our discussion could proceed either by describing these three structural components alternately for both sexes or by discussing the reproductive system of each sex separately. We shall follow the latter, more usual approach because it permits a more coherent description of the progress of the germ cell through each system and because of the practical considerations involved in relating text to illustration. To offset the risk of reinforcing the common attitude that the reproductive systems of the two sexes belong to different worlds, it should be pointed out that such duality makes impossible the proper understanding of sexual function in physical or psychological terms. The first and foremost objective of our discussion of anatomy should be to underscore the basic similarity as well as the complementary structure of the two sexes without glossing over important differences. Attempts at reconciling this apparent contradiction between the unity and the uniqueness of the two sexes will continue to preoccupy us throughout this book.

In the male, the first component of the tripartite genital model suggested above consists of a pair of testes (*Figure 1*). The word is derived from the root for "witness" (testify) based on the ancient custom of solemnly placing the hand on the genitals when taking an oath.

Each testis or testicle is enclosed in a tough fibrous sheath and suspended from a *spermatic cord* in a separate compartment of the scrotal sac. Each of these characteristics has important practical consequences. When the organ attempts to swell, for example during an infection, the unyielding cover will not give way but will choke its delicate structures. This condition, which occurs when an adult male develops mumps involving the testes, may result in sterility. Prepubescent boys are in no danger because their sperm-producing structures are not yet functional and thus not subject to damage. Females do not face this threat at any age because the ovaries, which are not enclosed in a tough sheath, are free to swell when similarly inflamed.

The fact that the testes are suspended within the scrotum means that, unlike the ovaries and most other organs of the human body, they are actually outside the body cavity. They are in the curious position of being "internal" organs that are external to the body. This

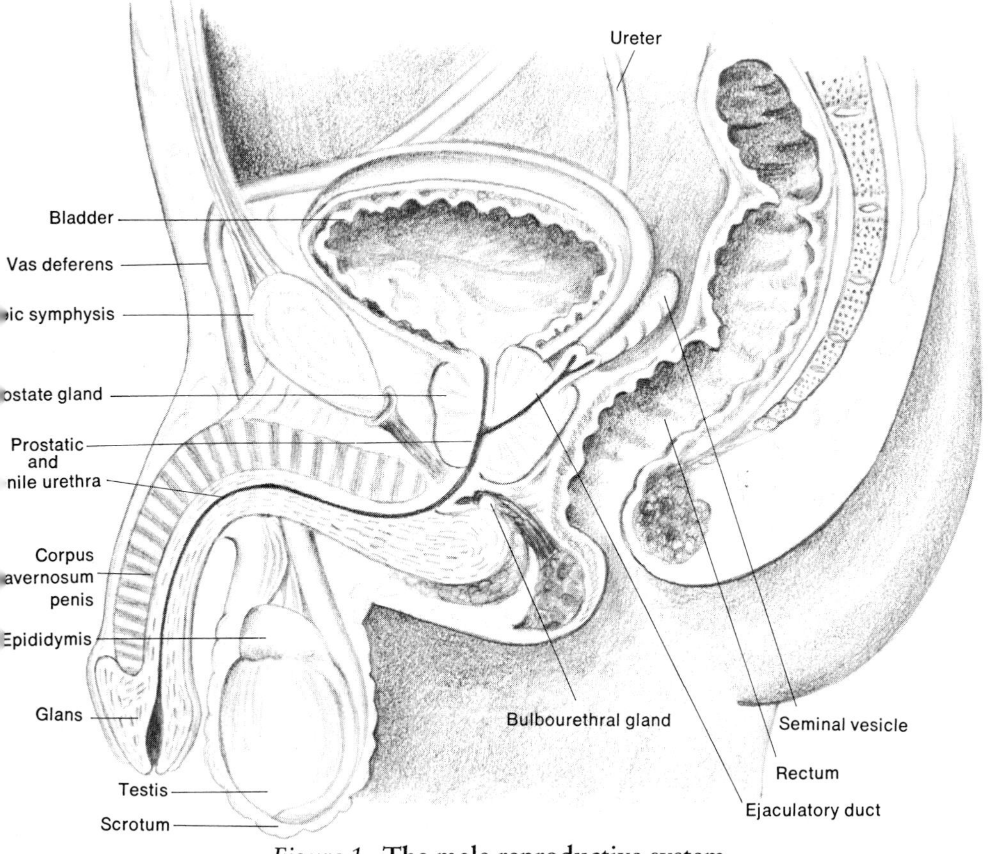

Figure 1. **The male reproductive system**

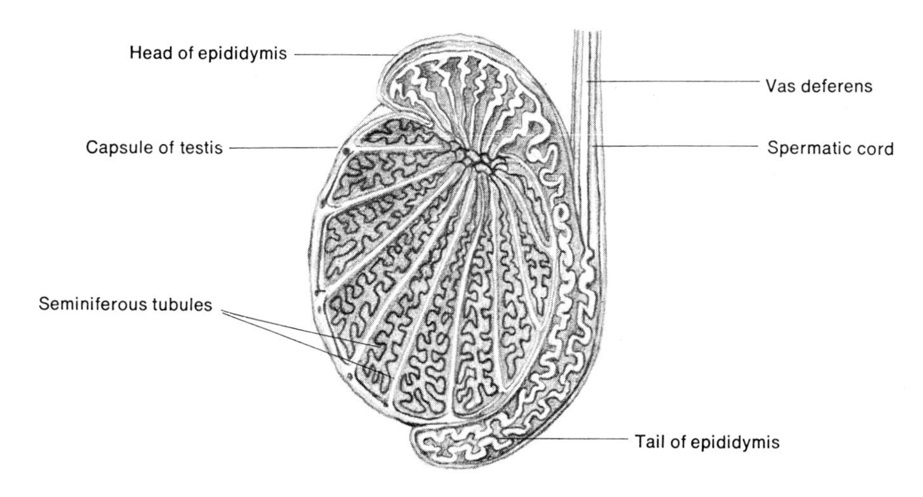

Figure 2. **Testis and epididymis**

roundabout arrangement, however, is no mere anatomic curiosity. Spermatogenesis, the production of sperm, is highly sensitive to temperature differences; it is hampered by the warm internal environment but will proceed optimally within the scrotal sac where temperatures are somewhat lower. The scrotal sac helps to maintain the temperature balance by relaxing into a baggy pouch when warm (thus exposing more surface to heat loss) and contracting into a tight wrinkled state in response to cold.

The testes develop high up in the abdominal cavity of the fetus and gradually descend to the edge of the pelvis. Usually they have descended into the scrotum by the eighth month of intrauterine life, but in a few cases they make this descent shortly after birth or in infancy. It is imperative that this occur prior to puberty because undescended testes are usually sterile. (Because hormonal functions are not affected in these cases, all sexual characteristics other than fertility will be intact.) Today the condition is usually detected in time and easily corrected medically or surgically. Incidentally, undescended testes are more likely to develop cancer. Another problem related to testicular descent is that the passage traversed by the testes may not become obliterated as it ought to (or it reopens when tissues become slack in old age), which permits intestinal loops to slide into the scrotal sac and results in the well-known condition called "rupture" or inguinal hernia.

If one were to study the internal structure of a testicle under magnification, it would be seen to consist of conical lobules packed with highly tortuous threadlike tubules (*Figure* 2). Sperm production takes place in these delicate structures, called *seminiferous* or sperm-bearing tubules, which comprise a vast network. Their combined length, measuring hundreds of feet, permits the production of countless billions of sperm during a male's fertile lifetime. The production of sex hormones is independent of these sperm-manufacturing structures. Cells responsible for hormone production, which are located between the seminiferous tubules and in proximity to blood vessels, are called *interstitial cells.*

The seminiferous tubules converge into a collecting network of ducts that ultimately empty into a single tube leading away from each testicle. The first portion of this tube (the *epididymis*) is highly tortuous but, because it is folded upon itself, it appears as a C-shaped structure clinging to the surface of the testicle. Its length, about twenty feet, provides a vast capacity for the storage of semen. The shorter and straighter continuation of the epididymis, known as the *vas deferens*, is one of the components of the spermatic cord, mentioned earlier, from which the testicle is suspended. During its upward course within the scrotum, the vas deferens can be felt as a firm cord before it disappears into the abdominal cavity. The fact that this structure is easily located and surgically accessible makes it the most convenient target for sterilizing men. The operation, known as *vasectomy*, simply involves the cutting or tying of the vas (on both sides) through two small incisions performed under local anesthesia. The procedure results in permanent sterility with no physiological effects whatsoever on sexual desire, performance, or male characteristics. Because sperm contribute so little to the volume of the semen, there is not even a noticeable effect upon the quantity of the ejaculate. In rare cases, fertility may be reestablished by resuturing the vas, but this is an exceedingly difficult and unreliable procedure. New techniques may improve the chances of reestablishing fertility for the man who has a change of heart.

Within the abdominal cavity, each vas deferens curves around the urinary bladder and is joined behind the bladder by a small structure called the *seminal vesicle* whose secretions are believed to activate the motility of sperm (*Figure 3*). The tip of the vas deferens joins the duct of the seminal vesicle to form the *ejaculatory duct*, which runs its entire length within the *prostate gland*, an organ the size and shape of a chestnut, located at the base of the urinary bladder. Through the prostate passes the *urethra*, the tube that leads out of the bladder (*Figure 4*). The urethra should not be confused with the *ureters* that convey urine from the kidneys to the urinary bladder. The ejaculatory duct on each side empties into the urethra as it traverses the prostate gland, and there on a single passage both urine and semen are conveyed to the outside—an anatomical feature characteristic of the male but not of the female, whose urethra conveys only urine and has no direct link with the reproductive system.

The prostate deserves additional mention because its secretions account for much of the volume of semen as well as its characteristic odor. These secretions are emptied into the urethra during ejaculation through a sievelike set of ducts. The prostate often enlarges in older men and then causes difficulty in urination. Cancer of the prostate occurs more frequently than any other malignant tumor found in men, and is more prevalent among men of advanced years. The Cowper's or *bulbourethral glands*, named for William Cowper who first described them in the seventeenth century, produce the clear, sticky fluid that appears at the tip of the penis during sexual arousal. This fluid should not be confused with semen although it sometimes contains stray sperm that can result in pregnancy even though intercourse has not culminated in ejaculation.

The male organ for sexual intercourse and for the delivery of semen

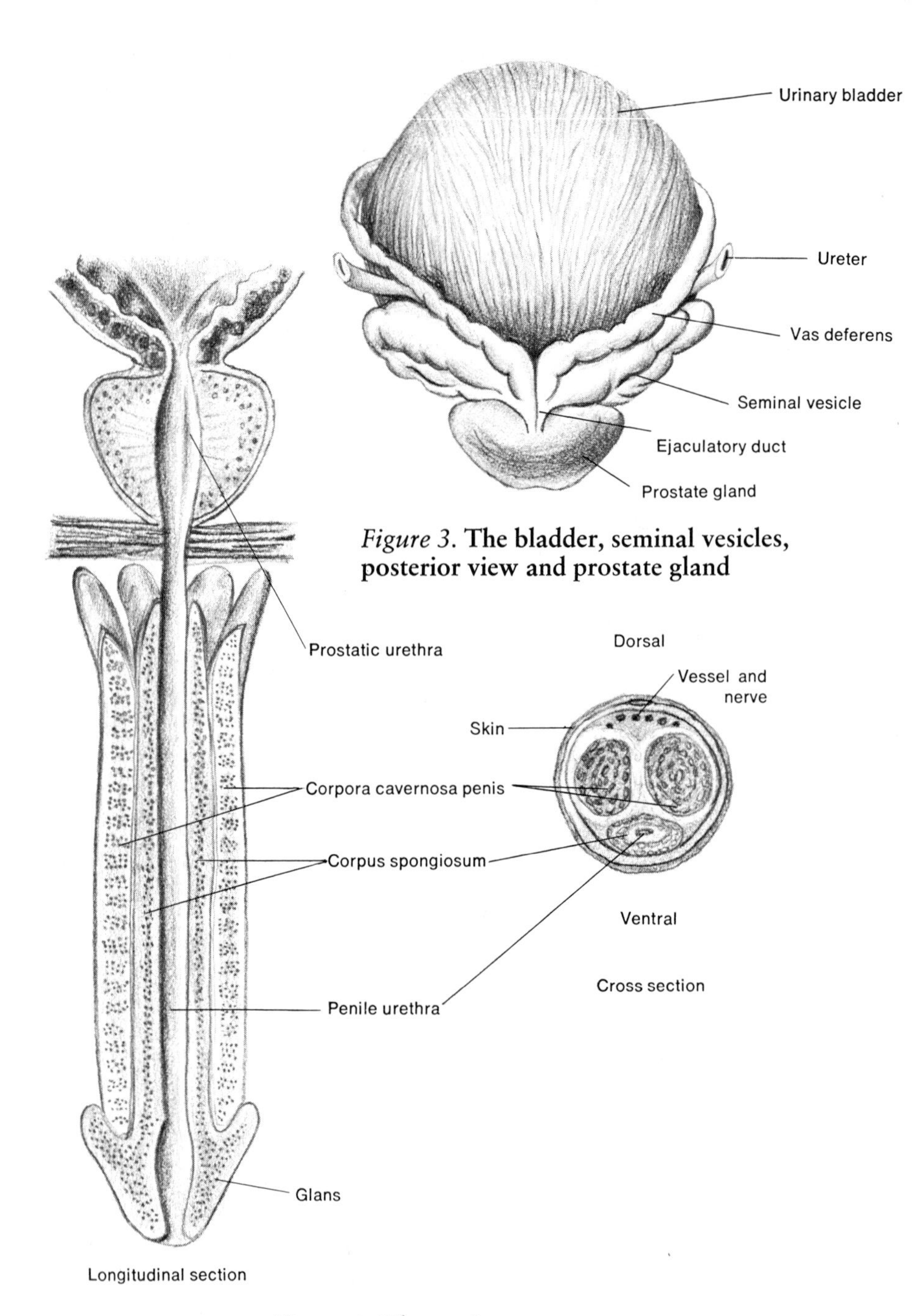

Figure 3. **The bladder, seminal vesicles, posterior view and prostate gland**

Figure 4. **The penis**

for reproductive purposes is, of course, the penis. It consists of three cylindrical bodies of spongy tissue, two *corpora cavernosa* and one *corpus spongiosum*; the urethra courses through the latter (*Figure* 4). The penis ends in an acornlike enlargement called the *glans penis*, which is highly endowed with nerves and constitutes the most sensitive part of the male sex organs. The penis has no bone and no intrinsic muscles. Erection is a purely vascular phenomenon. The components of the penis are sheathed in fibrous coats and enclosed within its loose skin. Normally this skin covers the glans but can be withdrawn easily. Circumcision, the surgical removal of the foreskin, leaves the glans permanently exposed.

The penis is the subject of a rich folklore. At various times and places it has been the object of religious veneration as a fertility symbol. In ancient Greece, phallic worship was associated with Priapus (son of Aphrodite, goddess of love, and Dionysus, god of fertility and wine). In India, Shiva or Siva, one of three supreme gods of Hinduism, was symbolized by an erect penis called the "lingam." Marriages were sometimes ritually consummated with stone phalluses in the hope of promoting fertility. Phallic worship has been prominent in certain Japanese and American Indian fertility rites.

While such practices have ceased to exist in their original forms, preoccupation with the penis continues. This pride and joy of many a man has come to be endowed with a variety of groundless attributes. Some of these fallacies are widely accepted and reinforced by "common sense" or selective personal experience. One such belief involves the size of the penis. A flaccid penis, usually three to four inches long, enlarges to about six inches when erect, while its diameter increases by ¼ inch to 1¼ inches. Perfectly adequate penises can be considerably smaller, and penises larger than 13 inches have been reported. The size and shape of the penis have very little to do with the competence of the man in giving or receiving sexual satisfaction. Size and shape are not related to his body build, skin color, or sexual experience, and are no more significant than the size and shape of his nose. In fact, whatever variation exists between flaccid penises tends to be reduced upon erection, since smaller penises tend to get proportionately larger than penises that were larger to start with. Current evidence also fails to substantiate long-standing allegations that a circumcised male is more rapidly aroused during coitus (and therefore quicker to ejaculate) than one who is not circumcised.

Because the female sex organs are built on the same basic plan, a comparative description is appropriate as long as it does not convey the impression that the female sex organs are significant only relative to those of the male. Female sexuality, anatomically and otherwise, needs to be understood in its own right. There are deeply rooted and not always conscious biases in this regard, and instances of such biases are reflected by the examples and metaphors we use to describe sexual organs and behavior. When we compare the penis and the vagina to a sword and its scabbard, for example, we are reducing the female organ to nothing more than an accessory, since a scabbard is useless without a sword. Of course the vagina is no more and no less useful or valuable than the penis. While such depreciatory comparisons are obviously silly, it is amazing to what extent these notions continue to influence our thinking in subtle ways.

The gonads in the female are the paired ovaries (*Figures 5* and *6*) that are homologous with the testes (arise from the same embryological origins) and fulfill a similar dual function, production of germ cells and sex hormones. The ovaries are smaller than the testes and, unlike the male gonads, remain within the abdominal cavity of the fetus. Microscopic study of the ovary during the fertile years shows that it contains numerous capsules or follicles in various stages of development (*Figure 7*). Within these follicles are the germ cells, the ova or eggs. One ovum matures each month during an ordinary ovarian cycle. Unlike the testes that produce billions of sperm during the postpubescent life of the male, the ovaries are believed to be endowed with their full complement of immature ova at birth; estimates of their number vary but are within the range of several hundred thousand. The monthly maturation of ova comes to an end during the menopause. (This process as well as the anatomy of the ovary are discussed further in connection with sex hormones.)

The tubal system of the female is much simpler and shorter than that of the male (*Figure 6*). A uterine tube about four inches long provides passage on each side from the ovary to the centrally located uterus. These tubes are named for the sixteenth century anatomist Gabriello Fallopio, who mistakenly thought they functioned as "ventilators" for the uterus!

The uterine or *Fallopian tubes* are attached to the sides of the uterus near its top, and their openings lead into the uterine cavity. Their ovarian ends, however, are entirely free; even though fringed projections keep them close to the ovaries, there is no tube leading out of the ovary. The mature follicle ruptures through the ovarian wall and releases the egg, which then must find its way into the funnel-like end of the Fallopian tube—despite the fact that the ovum is about the size of the tip of a needle, and the opening of the tube is a slit the size of a printed hyphen. Amazingly enough, there have been cases in which a

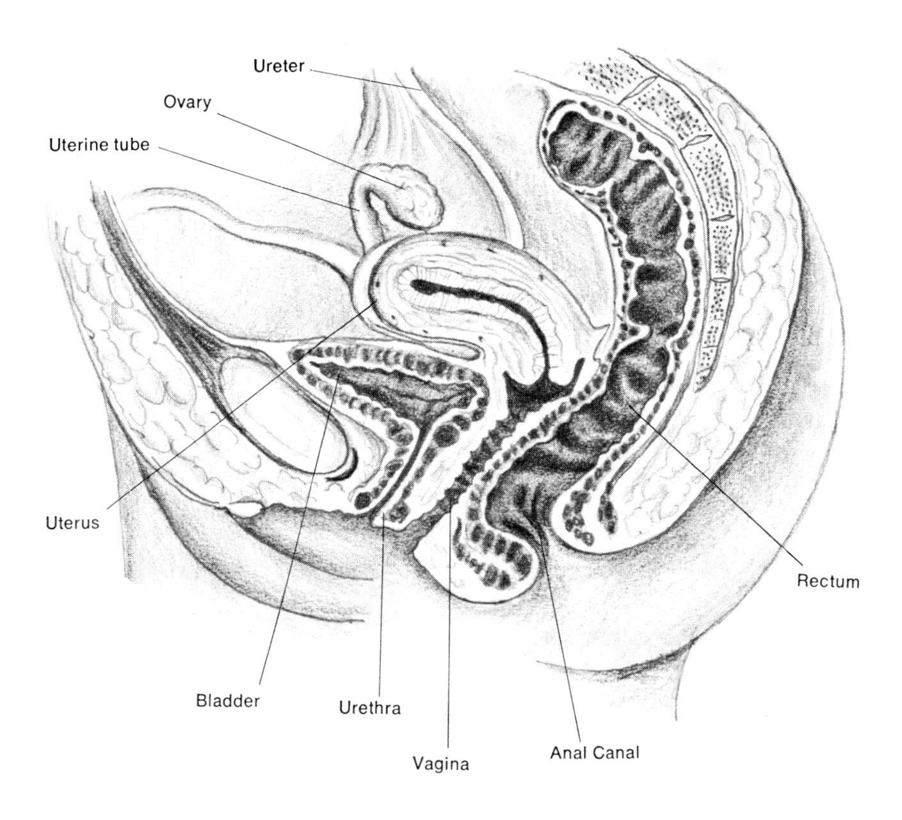

Figure 5. **Side view of the female pelvis**

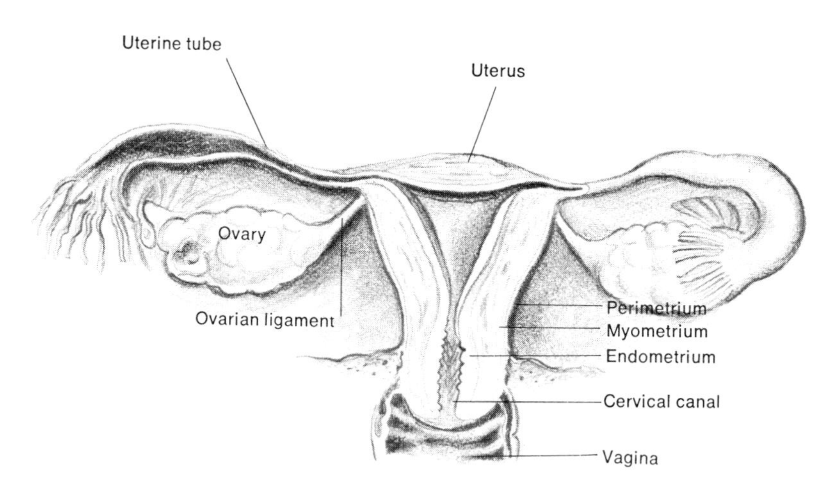

Figure 6. **Internal female reproductive organs**

woman missing a tube on one side and an ovary on the other side has nevertheless become pregnant.

The uterine tubes are the most frequently used targets for sterilizing women but, because they are not as accessible as the vas deferens, ligation of the tubes is a far more complicated procedure than vasectomy.

The uterine tube is not merely a passageway for ova; fertilization occurs in the tube rather than in the uterus or in the vagina. The fertilized egg, known as a *zygote*, then proceeds slowly to the uterine cavity. By the time it becomes embedded in the uterine wall, it is already a rapidly dividing cluster of cells on its way to becoming an embryo.

Occasionally the zygote becomes implanted in the wall of the uterine tube itself. This is one instance of an *ectopic* (out of place) pregnancy, which like all nonuterine pregnancies cannot come to full term. In a tubal pregnancy, sooner or later the tube ruptures, resulting in the death of the fetus and catastrophic consequences for the mother in the absence of prompt surgical intervention.

The uterus or womb is a pear-shaped, muscular organ that expands greatly during pregnancy. Normally it has a narrow cavity with a special lining that undergoes extensive changes during the menstrual cycle and in pregnancy. The lower portion of the uterus projects into the vagina and is known as the cervix. The "smear test" recommended to all women as a periodic check for cancer consists of examining scrapings from the cervical surface for abnormal cells.

The uterus and breast are frequently sites of cancer—one more compelling reason why the sex organs ought not to be shrouded in mystery and ignorance. Fortunately cancers of the uterus and of the breast give early warnings that, when heeded, may well make the difference between life and death. Not every lump or enlargement of the breast or swelling in the armpit constitutes breast cancer, and there are many benign causes for abnormal vaginal discharges and nonmenstrual bleeding; but these are some of the early signs of cancer, and women should be alert to them without becoming preoccupied by them.

So as not to leave the uterus with the grim associations raised above, let me mention an historical curiosity that has found its way into our language. The Greek word for uterus is "hystera," which appears in terms like *hysterectomy* (surgical removal of the uterus) and also *hysteria*, which the Greek physicians believed resulted from the uterus wandering in search of a child. Over twenty centuries later Freud was to return to this theme of the sexual component in hysterical conditions—perhaps in a sense vindicating the derivation of the term.

The vagina is the functional counterpart of the penis (but not its anatomical homologue) as the female organ for sexual intercourse and as the recipient of the male ejaculate. It is also the outlet for the menstrual flow (but not urine) and part of the canal that the baby traverses at birth. During some of these functions the vagina is capable of great expansion, but ordinarily it is a collapsed muscular pouch, a potential rather than a permanent space. Its main surfaces are formed by its front and back walls; its side walls are narrow, which is why the vaginal cavity looks like a slit slanted downward and forward in anatomical side views (*Figure 5*).

At its upper end the vagina surrounds the cervix, which opens into it. Its external opening, called the *introitus*, is flanked by elongated masses of erectile tissue and the *bulbocavernosus* muscle. Combined with other muscles in the region, these structures endow a woman with considerable control over her vaginal opening. Because it is not as direct or as effective as the control exercised, for instance, by the anal sphincter, many women may be hardly conscious of it; nevertheless this anatomical feature has important bearing on sexual functions.

The virginal introitus is partially covered by the hymen, a delicate pinkish membrane with no known physiological function (*Figure 8*). The size and shape of the hymen varies. It may surround or bridge the vaginal opening or form a sievelike cover for it. There must always be some opening to the outside. Rare cases in which there is no such opening, a condition known as *imperforate hymen*, are usually detected when menstruation starts and the flow accumulates month after month, causing the vagina and the uterus to swell. The condition is readily corrected by surgical perforation of the hymen.

Most hymens will permit passage of a finger or a sanitary tampon but cannot accommodate an erect penis without tearing. Since some hymens can withstand intercourse and others get torn accidentally in nonsexual activities, the presence or absence of an intact hymen does not constitute a reliable criterion of virginity.

Although defloration is no longer the great issue it used to be, some aspects of it continue to need clarification. It is not a formidable undertaking calling for heroic effort and fortitude. In the heat of sexual excitement, which hopefully accompanies the event, the woman feels minimal pain and bleeding is usually slight. It can nevertheless be a traumatic event. The torn structure, after all, is made of flesh and blood, not tissue paper. What is more important, an anxious, tense, and unstimulated woman will be likely to experience sustained muscular tension that causes pain, and clumsy attempts at forcefully penetrating an unlubricated vagina certainly do not facilitate matters. Although some women go as far as having their hymens surgically incised prior to their first experience, in most cases all that is needed is some knowledge of

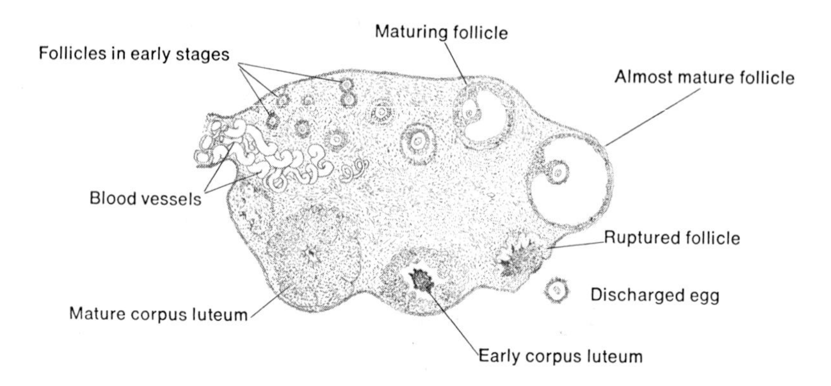

Figure 7. **Composite view of ovum**

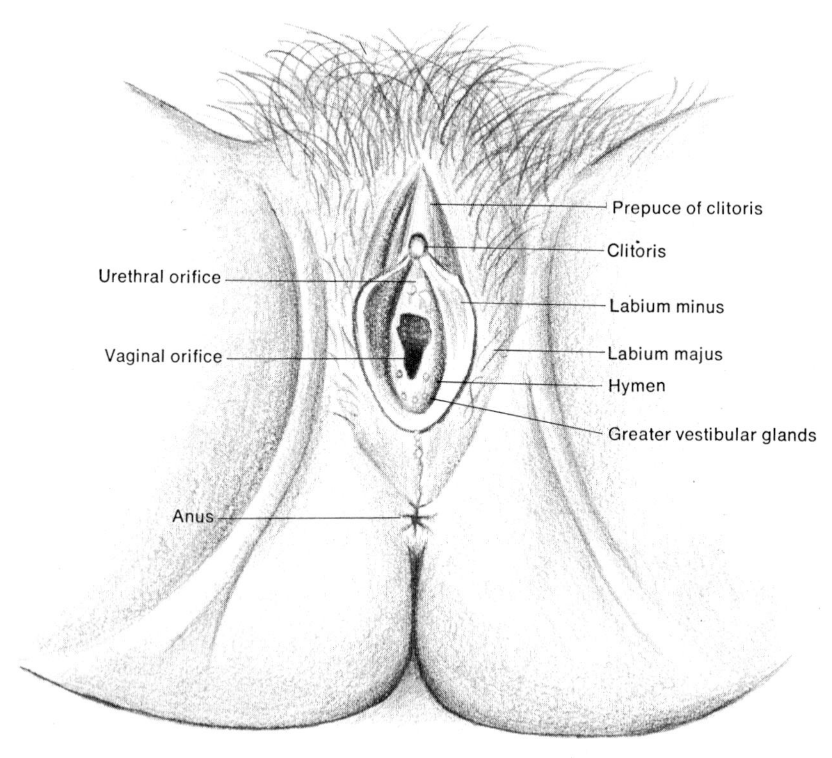

Figure 8. **External female genitalia**

what to do and what to expect and some consideration for the sexual partner.

Curiously enough, the hymen is an exclusively human structure. Other primates and lower animals do not have it, and its evolutionary significance is unclear; but many human societies have made the most of it. Many cultures have emphasized the integrity of the hymen at the time of marriage, and failure in this respect has constituted grounds for penalties ranging from disgrace to torture and death. Confronted with such prospects, some women undergo plastic surgery to "revirginize" themselves.

Another set of structures closely associated with the vaginal opening consists of a pair of glands (very small—even though they are called the *greater vestibular glands*) whose ducts open on the outside of the hymenal ridge. These glands were considered significant until the recent discovery that they fulfill only a minor and unessential role in vaginal lubrication.

Although the vagina cannot compete with the penis as a source of unwarranted claims and unfounded legends, it has its share of fallacies. It should be mentioned that the sources of this intelligence are almost always men, professional or otherwise. This is true concerning opinions about the penis as well. In the erotic literature one often reads about women waxing ecstatic about the glories of some gigantic penis, but these words are placed in female mouths by male authors. Women have generally tended to remain silent about these matters.

First let us consider the size of the vagina. In most women it is three to four inches long, not including the *labia* or "lips" flanking the opening (described below), which in a sense must also be traversed by the penis during coitus. These dimensions, however, are almost as meaningless as they would be if applied to a balloon, since the vagina functions somewhat like a balloon. Normally collapsed, it yields to accommodate, within wide limits, structures of various sizes. Since no penis is as large as a baby's head, which must move through the vagina at birth, for all practical purposes it is senseless to talk about vaginas that are "too small" or "too large." But some men will still insist that some vaginas "feel tight" and others "feel lax"—that is the way it is, and no anatomical argument will convince them otherwise. Some women also support the observation that the "fit" during coitus varies from one person to another, or even from one time to another.

This apparent contradiction can be resolved if one considers the properties of the introitus separately from the characteristics of the rest of the vagina. The inner portions of the vagina are readily distensible and its walls, being poorly supplied by nerves, are quite insensitive. But the introitus and the parts adjacent to it are a highly sensitive area that swells in response to stimulation, which changes both the size of the opening and the feel of the tissues. Furthermore, the state of tension of the various muscular components in the region is highly influenced by the level of sexual excitement and general state of mind of the woman. All these factors combine to create functional differences between different coital experiences. The difference then is not in fixed anatomic features but in their functional aspects, which are influenced tremendously by the psychological forces at play. How a woman feels to a man, even in a literal sense, is a function of how she herself feels. There is more to be said in this regard when we discuss the joys and sorrows of sexual intercourse.

Some men imagine all sorts of things about the vagina. Some are horrified by fantasies that the vagina is full of ground glass or razor blades, or that it is armed with teeth (*vagina dentata*). Under such presumed circumstances, when discretion understandably proves to be the better part of valor, these men become impotent—thus obviating the risk of life and limb in the heat of passion. True, these fears are pathological, but that does not make such feelings any less human.

Finally, there is a misconception that the penis can become "trapped" within the vagina. While some men could think of worse fates, others are made uneasy by the prospect. But no one need fear this predicament since it is physically impossible among humans. The inspiration for this notion comes from the observation of dogs that get "locked" in this way. A bone in the dog's penis permits a semi-erect penis to penetrate. On full erection, the tip of the canine penis expands into a knot that cannot be withdrawn before ejaculation and loss of erection.

The external genitalia of the female are known as the *vulva*, which means "covering," or the *pudendum*, which means "a thing of shame"! They include the major and minor lips, the clitoris, and the vaginal introitus, which is normally concealed from view. The most readily visible part of the female genitalia is the slight protuberance known as the *mons pubis* or *mons veneris* (mount of Venus), which becomes covered with pubic hair following puberty. The major lips or *labia majora* that curve downward between the legs vary in prominence. Between them are the hairless folds of the small and pinkish minor lips or *labia minora*. Within the space enclosed by the minor lips (the vestibule) are the vaginal orifice and, above it, the urethral opening through which the urinary bladder empties to the outside. As mentioned earlier, the female urethra is totally independent of the sexual organs. The inner lips at their upper end split into two layers that surround the *clitoris*, which is above the urethral opening. The sequence of structures within the vestibule from front to back is: clitoris, urethral opening, vaginal introitus. The anus lies further back and totally outside the vestibule.

The clitoris is a unique structure in that its sole purpose is sexual; it has no reproductive function. The clitoris is the homologue of the penis, and like the penis it consists of spongy, erectile tissue. It is not, however, a precise replica or a miniature penis. Profusely endowed with nerves, it functionally resembles the glans penis rather than the penis as a whole. Even though the clitoris swells during sexual excitement, it does not become erect because its overhanging *prepuce*, the upper layer of the labia minora, holds it down.

The so-called circumcision of the clitoris or *clitoridectomy*, formerly practiced in some cultures, was actually a mutilating procedure whereby the clitoris was amputated for ritualistic purposes. The practice has not been as widespread as male circumcision and has no redeeming justification. It is a good example of the peculiar practices that often characterize the sexual life of human beings.

Sex Hormones

Hormones, chemical substances secreted into the blood stream by the endocrine glands, have specific and far-reaching effects on other organs and tissues. Even though the concept of such remote chemical control can be traced to ancient notions of body humors, the modern science of endocrinology is still young. The term "hormone" (from the Greek word for "excite") was first used at the turn of this century. Since then more than twenty hormones have been discovered, and many of these have some bearing on sexual development and function. Hormones that play a central role in this regard are known as the sex hormones. Those that occur in higher concentrations in the male are known as the male sex hormones (*androgens*) and those that are more abundantly produced in the female are the female sex hormones (*estrogens* and *progesterone*).

The influence of the sex hormones, probably extending throughout the life span, includes some effects that are not even related to sex. The more critical and dramatic effects of sex hormones occur during the early development and differentiation of the reproductive system and then again during puberty. In both instances hormones are basically intermediary agents that bring about these changes. Hormone production and functions are controlled by the brain and ultimately by genetic factors. Now we must backtrack to the very beginning of life and trace these developments chronologically.

The genetic sex of the individual is decided at the moment of fertilization and is immutable. With the exception of sperm and ova, all human cells normally have twenty-three pairs of chromosomes. Twenty-two of these pairs look alike in both sexes. The twenty-third pair comprises the "sex chromosomes." In the female the sex chromosomes are similar and are labeled XX. The male pair, however, has one X chromosome and one Y chromosome. A sperm or an ovum has only half the number of chromosomes found in any other cell of the human body. When sperm and ovum unite, they recreate in the zygote the full complement of twenty-three pairs of chromosomes. Since the sex chromosomes in the female are similar, all ova are alike in the sense that each has one X chromosome (half of the XX pair). In males, however, since an asymmetrical twenty-third pair is being split, some sperm will carry an X chromosome and others a Y chromosome. If an X chromosome-bearing sperm joins an ovum (which already has an X chromosome), the result will be a female zygote (with XX sex chromosomes). If a sperm bearing a Y chromosome is involved, an XY configuration is created and the issue becomes male. In humans, therefore, the sex of the child is always determined by the father.

It is the sex chromosomes that carry the genes with the coded instructions that will direct the sexual differentiation of the individual. Although we do not yet fully understand how these exquisitely complex operations are carried out, we know that in one fashion or another the sex hormones play crucial roles in mediating this process. We know also that these influences occur at critical times; it is not only a question of *what* hormones and *how much* of each is necessary, but also *when* a given concentration of a hormone needs to be present for sexual development to proceed normally.

The genital system of the embryo appears during the fifth or sixth week of intrauterine life. At this stage the reproductive system is as yet undifferentiated. Both sexes have a pair of gonads and two sets of ducts that look alike. This is not to say that the sex of the child is still undetermined; as stated earlier, that issue was settled at the moment of fertilization. But now for a few weeks the sexual development follows an identical pattern in both sexes. This is important to keep in mind because inferences are sometimes drawn from this biological characteristic whereby every person is claimed to be fundamentally "bisexual." This is not the case. At this early stage in life, the gonad is not a testis as well as an ovary; it is as yet neither a testis nor an ovary but will become one or the other depending upon the genetic sex of the individual. There are very rare instances in which an individual is born with both testes and ovaries (true hermaphrodites), but these are pathological exceptions to the rule.

The undifferentiated phase of genital development does not last very long. By the seventh week of intrauterine life the gonad starts to show signs of differentiation, and by the tenth week it is clear whether the organ is destined to become a testis or an ovary. The gonads do not attain full maturity until puberty. The progressive differentiation of the genital ducts and the external genitalia follow the gonadal changes. By the fourth month the sex of the fetus is unmistakably clear by external appearance alone.

What has been described so far is the normal course of events. Because of genetic abnormalities or hormonal defects, peculiar and paradoxical developmental problems may arise. For example, the sex chromosomes, instead of consisting of the normal XX or XY pairs, may lack a chromosome (XO) or may have an extra chromosome (XXX, XXY, or XYY). On the other hand, a normal chromosomal pattern by itself does not guarantee normal sexual development if the intermediary hormonal mechanisms are disturbed or lacking. A classical example is castration. If the testicles of a genetically normal prepubescent boy were removed, the main source of androgens would be eliminated, and the boy would fail to mature sexually. Under these circumstances, the genetic pattern could not assert itself any more than a conductor could make music without an orchestra.

Such is the power of hormones during key periods of life that "reversal" of sexes is feasible in animals in the laboratory. By removal and transplantation of gonads in newborn rats (exchanging testes with ovaries) genetically male or female animals can be made to function physiologically in a manner characteristic of the other sex. Much of what we have learned about the effect of hormones comes from animal studies, in addition to the natural "experiments" that occur in humans when something goes wrong. This type of serious genetic or hormonal abnormality is rare, however, and one cannot help but marvel that such intricate, delicately balanced, and integrated processes exist with so few errors.

Even more fascinating than the effects of hormones on the reproductive system is their possible influence on the brain and hence on behavior. Unfortunately there is as yet limited information on this subject, especially as it pertains to humans. Seymour Levine (of Stanford) has demonstrated how the early administration of hormones in newborn rats can radically reverse their subsequent sexual behavior. Because these animals fail to regain their sex-appropriate behavior even when they are later treated with hormones of their own sex, it seems that exposure to these substances at a critical time leaves an indelible impression on the brain. Julian M. Davidson (also at Stanford) has extended these observations further, showing the effect of hormones on the rat brain by implanting hormones in the brain tissue itself. These findings cannot be extended automatically to humans, of course, nor can such experiments be replicated on people. We are as yet unsure whether or how sex hormones affect the brain differentially in the two sexes. If it could be shown that the brains of males and of females are "wired" differently, the implications would be far-reaching. Although many persons now assume this to be the case, there is no solid evidence to support the contention as yet.

If embryonal hormones have performed their tasks adequately, the infant will be born with a clearly differentiated but still immature male or female reproductive system. Although hormones continue to exist in low concentrations throughout childhood, they do not come into their own again until puberty, when under their influence the definitive adult reproductive system becomes established.

During puberty a part of the brain called the *hypothalamus* triggers into increased action the front portion of a pea-size gland called the *pituitary*, actually an extension of the brain projecting out of its lower surface (*Figure 9*). In response, the pubescent female pituitary produces the *follicle-stimulating hormone* (FSH) that initiates the maturation of ovarian follicles and prompts the ovary itself to produce *estrogen* (a term that actually stands for a class of compounds), the first of two female hormones. When the estrogen level in the blood reaches a certain concentration it triggers the release of the second pituitary hormone called the *luteinizing hormone* (LH). Under the effect of LH, the mature ovum is released and literally bursts out of the ovary. The empty follicle now develops into the *corpus luteum* or "yellow body" and under the continued influence of LH it produces in its own turn the second of the female hormones, *progesterone*, as well as more estrogen. Acting on the same feedback model, increased levels of progesterone induce the hypothalamus to inhibit the production of LH by the pituitary. The cycle is now complete. Since estrogen and progestrone are used up in the body, their levels gradually fall to a point at which they no longer inhibit the hypothalamus, which in turn permits the pituitary to increase the production of FSH and a new cycle is begun.

Several tasks are accomplished through this basic female hormonal cycle. During puberty, under the influence of estrogen, the prepubescent girl gradually turns into a woman; the contours of her body change, her breasts enlarge, her genital organs develop more fully. Gradually, after some erratic starts and stops, she also starts to menstruate. It usually takes several years, however, from the onset of these changes until she is a fully fertile and sexually mature woman.

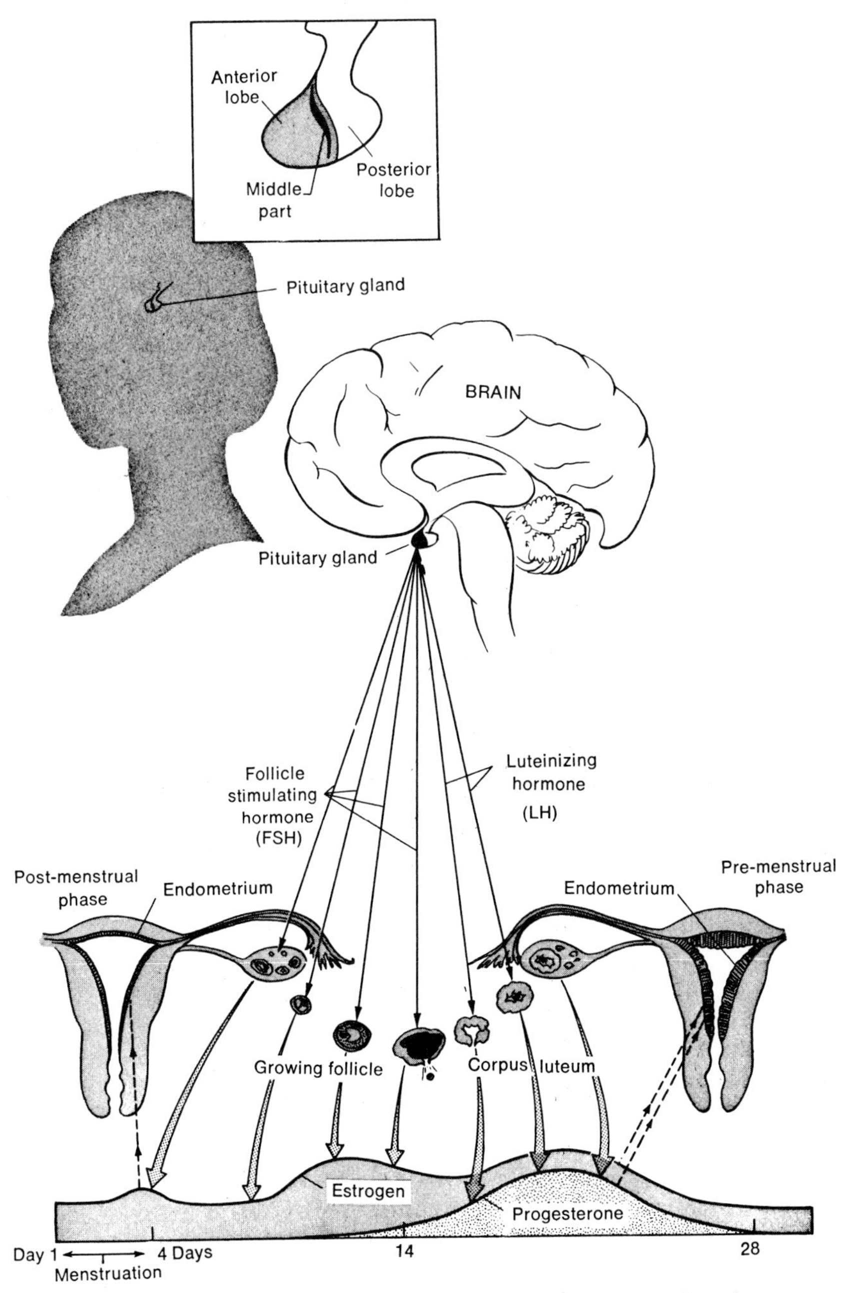

Figure 9. **Hormonal regulation of the menstrual cycle**

During her reproductively active years, a woman's monthly period is the reminder that this hormonal cycle is silently in progress. As you will note in *Figure 9*, on the left, the *endometrium* or uterine lining is quite thin immediately after a menstrual period. As the level of estrogen rises through increased production by the growing follicle, the lining of the uterus is gradually built up under its influence. Additional changes occur in the uterine lining with the continued effect of estrogen and the complementary influence of progesterone. Meanwhile the ovum has left the ovary. If it is not fertilized and implanted in the uterine wall, the levels of estrogen and progesterone drop, and the uterine lining is shed with some consequent bleeding—in other words, the woman menstruates. If a fertilized ovum becomes implanted and survives, new hormonal mechanisms go into effect and prevent menstruation from occurring.

At the time of the menopause this cycle becomes irregular and finally stops. The woman does not ovulate any more, and a drastic reduction in production of female hormones results in the slight "defeminization" of the postmenopausal woman, characterized by changes in her voice and growth of facial hair. These changes are due to small amounts of male hormones—produced by the adrenal glands common to both sexes—that normally exist in women. (The growth of pubic and axillary hair at puberty is also caused by androgens.) Prior to the menopause, the masculinizing effects of the male hormone are counteracted by the feminizing effects of estrogen. When estrogen production stops, androgens act unopposed.

Incidentally, birth control pills are simply various combinations of synthetic estrogen and progesterone-like compounds. Their action can be readily understood in reference to *Figure 9*. When a woman is taking the pill, a high level of these hormones is maintained in the blood stream. Normally this would occur when a follicle came to maturity and FSH production was therefore turned off. Since the pituitary cannot differentiate between the hormones of the ovary and that of the pill, it is "fooled" into repeating its customary response: with FSH turned off, new follicles do not mature, no egg is released, and hence no pregnancy can occur. When the woman stops taking the pill, she menstruates precisely for the same reason she normally does; in other words, a cycle with the pill is like a natural cycle except that it is anovulatory.

The hormonal pattern in the male is relatively simple. At puberty the hypothalamus prompts the anterior pituitary to produce the same two hormones described for the female: FSH and LH. But since the term "luteinizing" would be meaningless in the male, and because this hormone in the male acts upon the interstitial cells of the testes, it is known as the *interstitial-cell-stimulating hormone* (mercifully abbreviated to ICSH). Sperm production is initiated under the influence of FSH. The interstitial or Leydig cells are concurrently stimulated by ICSH to produce the male hormone or androgens of which *testosterone* is the primary component. Testosterone in turn initiates the various changes of puberty: change of voice, growth of facial, axillary, and pubic hair, further development of internal and external sex organs. It is after the full development of the prostate gland and other accessory structures that male orgasm culminates in ejaculation.

In the male as in the female, puberty extends over several years from its inception until full sexual and reproductive maturity is attained. Unlike the female pattern, however, the male hormone pattern is not cyclic. Production of sperm and of testosterone are more or less continuous and, although these activities decline with age, they never cease during the natural life of a male. Although the testes also produce small amounts of estrogen, the female hormone normally does not have an opportunity to show its effects because androgen production is sustained. When estrogen production is greatly increased, either by pathological conditions or by estrogen therapy, definite feminizing changes become apparent.

What about behavioral effects of sex hormones? Among animals hormones clearly control sexual behavior. The female becomes receptive at the time of ovulation and increased hormonal output during the estrous cycle. A bitch in heat will drive the male dogs in the neighborhood (as well as the neighbors) out of their minds. At other times there is no comparable interest in sex.

Although numerous attempts have been made to link sexual interest to phases of the menstrual cycle, no convincing relationship has been established; there is too much variation among women. Once sexual maturity is reached, we seem to be liberated from the control of hormones. Postmenopausal women continue to be interested in sex, and estrogen does not enhance female sexual desire—although paradoxically androgens do. Yet extra androgens do not seem to have a consistent effect on ordinary male libido.

Nor has it been possible as yet to link hormones with sexual orientation in behavioral terms; there is currently no reliable way of separating heterosexuals from homosexuals or to make any other group distinctions by endocrinological comparisons. This is not to say that such differences do not in fact exist. Practical methods of male hormone assessment are only now being refined. We probably have not heard the ending of this story yet, and whatever the answers turn out to be, they are not going to be simple ones.

Physiology of Sexual Functions

Even relative to the study of sex hormones, which is a new field, there has been very little research into the physiology of sexual functions, and standard physiology texts have little to say on this subject. It is common knowledge, of course, that an episode of sexual discharge starts with a period of arousal, which is released by orgasm followed by feelings of lassitude and satiety. The physiological and behavioral details of this process were not considered fit subjects for study, however, prior to the recent research by Masters and Johnson.

It is customary to consider the phase of excitement or tumescence separately from the process of orgasm or detumescence. Masters and Johnson have separated the sexual cycle into four phases: excitement, plateau (actually a period of sustained excitement), orgasm, and resolution. Although this division facilitates description, it should be remembered that the entire process is one integrated phenomenon.

Sexual Stimulation

As we all know, people can become sexually aroused spontaneously through their own thoughts, or in response to external stimulation, which may involve a variety of sights, sounds, smells, and tactile sensations. All sensory modalities can be involved in sexual arousal and usually are. For man, however, the most important is touch, the only modality that could be considered to operate on a reflexive basis: even a man in a coma can be brought to an erection by the stimulation of his genitals or inner thighs.

It is tempting to try to isolate the "basic" physiological means of sexual stimulation from the learned responses, but such attempts usually lead to a mechanistic approach whereby one is constantly searching for the correct buttons and levers to push and manipulate. Much of the "erogenous zones" literature is based on such simplistic assumptions.

For our purposes here, the pertinent information on sexual arousal can be summarized as follows: Some surfaces of the body are more highly innervated than others and therefore more sensitive. These would include the glans penis (but not the penile shaft), the clitoris, the minor lips, the vaginal introitus (but not the vaginal canal), the nipples, lips of the mouth, and fingertips. Genital organs, such as the glans penis or the clitoris, and their adjacent areas are sexually more sensitive than comparably innervated regions like the fingertips. The erotic potential of any surface or modality of sensation is modified by learning and past experience. There are no special nerves that carry sexual impulses any more than there are telephone cables that carry only glad news. The sensations received by the brain when one's thigh is stroked must be evaluated and given meaning. So important are psychological factors in this regard that an individual may react completely counter to the biological givens. It is possible for a woman to respond to clitoral stimulation with revulsion but to be aroused by the caressing of her neck or face—far less sensitive regions. Some women have reportedly reached orgasm by having their eyelashes stroked or pressure applied to their teeth.

In light of the above, the notion of erogenous zones is meaningful or meaningless depending upon what is meant by it. If we are thinking of clear-cut regions that are innervated by particular nerves and automatically respond to tactile stimulation, then such regions do not exist. But if we are referring to general and ill-defined regions that are highly innervated and culturally defined as being erotically potent, then the concept is useful in that we can expect most people to respond to their appropriate stimulation. Even questions of propriety are pertinent in this regard. For example, even though the anal region is highly sensitive, if a person reacts with revulsion to its stimulation, then it becomes effectively anti-erotic. The effective lover of either sex will have to determine the particular erogenous map of the partner.

Erotic stimulation through nontactile modalities operates through psychological means. So far as we know, there are no sights, sounds, smells, or tastes that are inherently erotic. Unlike the sense of touch, the other senses cannot be labeled erotic on a demonstrable reflexive basis: there are no smells, for example, that will elicit erection in a comatose man. Obviously we can be highly aroused by what we see, hear, or smell, but these reactions are all learned and therefore highly varied among individuals as well as among various cultures.

Smell plays a major role in influencing behavior among lower animals. Through chemicals excreted externally, animals attract, repel, or otherwise attempt to regulate the behavior of other organisms. These chemicals are called *pheromones*. There is some evidence that sex pheromones exist in infrahuman primates and possibly in man, but it is not yet conclusive. Cosmetic manufacturers, therefore, will have to continue to rely on their imaginations to bolster the aphrodisiac potential of their scents.

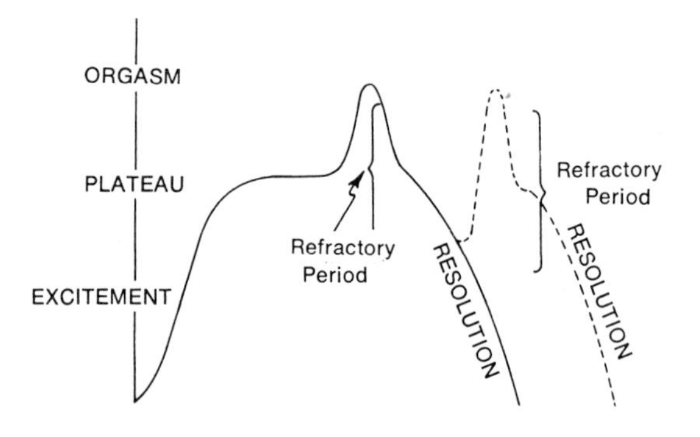

Figure 10. **The male sexual response cycle**

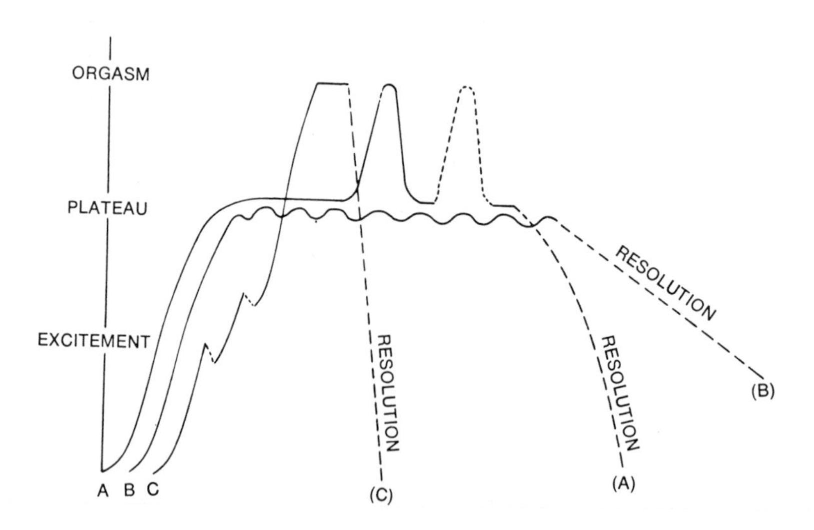

Figure 11. **The female sexual response cycle**

Sexual Response

Much of the information on response to effective sexual stimulation comes from the work of Masters and Johnson. Since this material is widely available in intelligible form, I will summarize the salient findings of this research and refer you to other sources for details.

At the risk of belaboring the point, I must reiterate the fact that sexual response, like sexual stimulation, involves the whole body; sex is in no sense a private affair between the genitals. The second cardinal rule is that, while we all share certain basic physiological characteristics, all these biological givens are subject to psychological modification. And while some generalities encompass all of us as human organisms, ultimately we can only be understood as individuals.

The sexual response cycle, as indicated earlier, is divided by Masters and Johnson into four phases: excitement, plateau, orgasm, and resolution (*Figures 10* and *11*). While individuals vary in regard to the length and intensity of these phases, a basic pattern holds true in general. Comparison between the male (*Figure 10*) and female (*Figure 11*) patterns nevertheless shows two important differences between the sexes. First, three models are required to represent the range of sexual response in women, who show much more variability than men. Pattern A in *Figure 11* is basically similar to the male sequence. In alternative B the plateau and orgasmic phases become fused, and the woman experiences a sustained series of orgasms. In pattern C excitement culminates in orgasm without a plateau phase. The second major sex difference is that the male orgasm is followed by a refractory period, an obligatory time span that must pass before the male can reexperience orgasm. Females can have multiple orgasms in succession because they have no refractory period.

The physiological responses of the body to sexual stimulation are in part general and identical in the two sexes and in part specific in each sex. General body reactions include nipple erection in the excitement phase (much more marked in women) and some flushing of the skin. In the plateau phase this flushing becomes more generalized and is accompanied by generalized muscular tension. The heart beats faster and the rate of breathing increases. During orgasm there are muscular contractions including the typical motions of pelvic thrusting. Tachycardia and hyperventilation persist, but return to normal during the resolution phase, when the individual also often perspires.

Erection of the penis, usually the first sign of sexual arousal in the male, may occur in a few seconds. There is also contraction of the scrotal sac and elevation of the testes within it. Erection occurs reflexively through the spinal erection center. It is a vascular phenomenon; although several mechanisms are involved, basically it results because the flow of blood into the penis exceeds outflow. The swelling of spongy tissues results in the erection and rigidity of the penis. The process is reversed when excitement abates and outflow exceeds inflow of blood until a balance is reestablished.

During the plateau phase the penis undergoes further engorgement, and the testes are pulled up within the scrotal sac—which for some reason makes orgasm predictably certain. The mucoid secretion of Cowper's glands appears. Orgasm in the postpubescent male is accompanied by ejaculation, another reflex process that starts with the contractions of the vas deferens, seminal vesicles, ejaculatory duct, and prostate, followed by penile contractions. In the resolution phase, structures return to their nonstimulated state.

In the female, vaginal lubrication is the first physical response in the excitement phase and may occur within 10 to 30 seconds of effective stimulation. The vaginal walls appear to be "perspiring." Other physiological responses are the congestion and thickening of the vaginal walls and the labia, expansion of the inner end of the vaginal canal, and elevation of the uterus. The clitoris becomes tumescent relatively late in this sequence.

During the plateau phase the outer third of the vagina swells markedly, while its inner two-thirds expand more fully, and the uterus is further elevated. An important change at this stage is the intense reddening of the minor lips, which indicates the inevitability of orgasm. A variable amount of mucoid secretion is produced by the bulbourethral glands. The clitoris shows a peculiar reaction at this point; without losing its congested state, the clitoris withdraws under its prepuce where it remains hidden for the rest of the sexual cycle until the resolution period.

Female orgasm is identical to that of the male except that it does not involve ejaculation. Contractions at 0.8 second intervals involve the uterus and especially the outer third of the vagina (called the *orgasmic platform*). Organs return to their nonstimulated state during the resolution phase.

All of these general as well as genital reactions can be explained on the basis of two fundamental physiological mechanisms: vasocongestion and increased muscular tension. The erection of the penis, swelling of the labia, and flushing of the skin are examples of the former. The orgasmic contractions are the culmination and discharge of muscular tension. Vasocongestion is beyond volitional control, but muscular tension is subject to the will. Judicious and timely use of muscular tensions are crucial for appropriate mounting of sexual excitement. Some women fail to reach orgasm because of their passivity and failure to bring large muscular masses into play by active participation in coitus.

There are many other reactions in addition to those listed above, including increased salivation, decreased sensitivity to pain, and suppression of the gagging reflex. Some people may feel numb or itch. The subjective sensations are variable. Orgasm is generally experienced as highly pleasurable, although its intensity varies among individuals or from one time to another, and it is often difficult to account for these differences.

Pleasure is usually discussed in connection with emotions and psychological phenomena. This emphasis sometimes leaves the impression that such phenomena are totally independent of physical considerations, as if they exist "in thin air." Obviously this is not the case; every thought and emotion has some neurophysiological representation in the brain. This fact has been illustrated by animal studies.

It has been shown, for example, that "pleasure centers" exist in the brain of the rat. When rats are experimentally "wired" in such a way that they stimulate these centers electrically by pressing a lever, they will continue this activity to the point of exhaustion. Similar research on monkeys indicates that such findings are not limited to the rat and may be generalizable ultimately to man. One of the leaders in the investigation of the psychophysiology of emotion is Karl Pribram of Stanford.

CHAPTER TWO

SEXUAL BEHAVIOR

QUESTIONS ABOUT the frequency of sexual activity are often raised in a variety of contexts. Although most educated parents now recognize that masturbation is an acceptable and harmless activity, many are concerned (as are the adolescents themselves) that masturbation may be harmful if "excessive." Disagreements arise between spouses about the frequency of making love. Even though nothing may be said about it openly, a couple may find it difficult to determine whether one of them is making unreasonable demands or the other is being unduly restrictive.

When Kinsey set out to quantify human sexual behavior, he had to determine what to count. He chose orgasm as his unit because it is the most discrete and unmistakably sexual event. He realized of course that there is a vast amount of sexual activity that never culminates in orgasm, but it would have been impossible to count and catalogue every fleeting sexual fantasy or stirring of the loins. The Kinsey study became primarily an investigation of orgasmic behavior, just as several decades later the Masters and Johnson study would become an investigation of the physiology of orgasm. A common criticism of both of these approaches is that they reduce sex to orgasm. Obviously there is more to sex than orgasm, but systematic study requires that problems be considered in manageable bits.

Duly impressed with the many facets of human sexual behavior and unimpressed by the contention that decent people only engage in coitus, Kinsey and his associates attempted to assess the "total outlets" of their subjects by adding up all orgasms attained by an individual, through whatever means, over an average week. Considered in this restrictive sense, the sample population as a whole was shown to attain orgasms through six means: sexual intercourse, masturbation, petting, homosexual contacts, sex dreams, and animal contacts. All other sexual activities, such as voyeurism or sadomasochism, contributed a negligible number of orgasms to the total outlet, although such behavior could also accompany the major outlets.

These six ways of attaining orgasm do not of course occur with equal frequency. Coitus and masturbation are, for example; incomparably more common than orgasm through sex dreams or animal contact. Although not everyone attains orgasms through all six means, those who rely on a single mode throughout their lives are also very rare. Most people attain orgasms predominantly by one means, supplemented by one or more additional outlets, depending upon such factors as marital status or life circumstances.

If you have already reacted with some discomfort at this attempt at sexual bookkeeping, your response is justified. To talk of total outlets conveys a rather simple-minded and mechanistic picture of sex. Chalking up orgasms in this fashion seems to disregard even the most obvious psychological and moral issues. The notion easily becomes comical if pushed far enough. Imagine extending the tally to families, organizations, even entire nations. We could, I suppose, even talk about our sexual GNP! But despite these shortcomings, the concept of total outlets has merit. Besides enabling researchers to quantify sexual behavior, it provides a fluid and dynamic view of sexual activity.

We are often tempted to think of individuals in discrete and opposing sexual categories, such as heterosexuals or homosexuals, masturbators or nonmasturbators, sadists or masochists. The available evidence indicates that such sharp distinctions are misleading. "The world is not to be divided into sheep and goats," wrote Kinsey. "Not all things are black nor all things white. It is a fundamental of taxonomy that nature rarely deals with discrete categories. Only the human mind invents categories and tries to force facts into separate pigeon-holes. The living world is a continuum in each and every one of its aspects. The sooner we learn this concerning human sexual behavior the sooner we shall reach a sound understanding of the realities of sex." (*Sexual Behavior in the Male*, p. 639.)

Instead of labeling individuals as heterosexual or homosexual, for example, it is often preferable to consider people with varying degrees of such experience (including *no* such experience). Sexual acts require labels, but when labels are applied to people, it tends to dehumanize them. It does *not* follow from this that everybody does everything (or would like to) if given the chance. Nor does it follow that all sexual behaviors are equally adaptive, healthy, edifying, or moral.

To return to our opening question—"How often?"—Kinsey found not one but many answers. There was so much variation in the total sexual outlets of individuals that it became necessary to examine the sample population in terms of age, sex, marital status, educational level, and other variables that were likely to be related significantly to frequency of outlet. Even within such subgroupings Kinsey's interviewers encountered wide differences. In this context, to say that among males from adolescence to age eighty-five there are on the average (mean) nearly three orgasms a week per person is meaningful as a crude estimate. But the average figure takes on a different meaning if one realizes that in this same population there would be one man with one orgasm in twenty years and another with twenty orgasms a week for twenty years (and both apparently healthy).

To assess frequencies of total outlet more intelligently, we need to examine the pertinent subgroups. The age of the individual was found by Kinsey to be the single most important factor related to the magnitude of total outlet, and there were significant differences between the two sexes in this regard. Among males total outlets attained their peak shortly after puberty and were maintained at this level until about age thirty. During this period there were about three orgasms (mean) in the group in a week with each average individual attaining two orgasms (median) during this period. (The apparent discrepancy between these two "averages" results from their statistical derivation.) After age thirty the total outlet steadily declined. By age fifty the average male was attaining one orgasm (median) a week; by age seventy-five he was attaining orgasm very infrequently. These averages, however, pertained to groups and not to individuals. There were some men in the seventy-five-year group whose total outlets exceeded those of many in the thirty-year group.

The corresponding rates for females showed some intriguing differences. Unlike the male pattern, the total outlets of females started very low and only gradually built up to a peak around age thirty. These rates were then maintained for a full decade up to about age forty, when they began to decline. Age for age, female rates were consistently lower than male rates. At the peak age of thirty-five the average (median) woman, for instance, was reaching barely one orgasm a week. But there were individual women even in the lowest total outlet group who were outdoing some men in the highest total outlet groups.

How do we explain these wide differences in total outlet between individuals and various subgroups? Some questions are relatively easy to answer. In a society in which marital coitus was considered the only acceptable sexual outlet for women, one would expect higher rates among wives than among single women, which is in fact what Kinsey found. Another possible factor is preselection, whereby women more interested in sex are more likely to get married. Kinsey also found that over a third (36 percent) of the women interviewed had not experienced orgasm prior to marriage; by contrast most men had had hundreds upon hundreds of orgasms by then. Evidently social factors are operating here also, but it may be naïve to explain it on that basis alone.

We are still ignorant of the precise factors that determine levels of total outlet. As one examines the available data, it becomes evident that a variety of biological, psychological, and cultural forces are at play in a highly interrelated fashion. Kinsey attempted to relate total sexual outlet to decade of birth, educational level, parental occupation class, religious background, and a rural-urban factor. The data provide some interesting comparisons but do not answer the question of causal relationships.

Ignorance of the roots of sexual activity has not prevented us from being fascinated by men and women with exceptionally active sex lives. The satyrs and nymphs of classical mythology have provided the labels for sexual "hyperactivity" among men (satyriasis) and women (nymphomania). Venetian adventurer Giovanni Giacomo Casanova de Seingalt (1725-1798), the fictional Don Juan, Catherine II of Russia (1729-1796), and assorted emperors and empresses are well-known representatives of this breed. While we hear of the exploits of the high and the mighty, no one keeps track of the staggering records that some happily married commonfolk may be setting in the anonymity of their bedrooms. At the other extreme are the sexually "underactive" people, a less celebrated lot who neither inspire artists nor keep moralists awake at night.

Are the sexually overactive especially gifted by nature—or are they sick? Are the sexually underactive blessed with a lust-free internal calm, or are they biologically defective or paralyzed by inhibitions? Clinicians have generally maintained that both extremes of sexual activity are pathological. There are countless case histories of neurotic people who frantically pursue sex or shun it altogether. Because the sexual behavior of these persons can be unmasked to reveal its neurotic nature, the clinicians argue, those in the population-at-large behaving similarly must be neurotic. But this assumption has not been proved; therefore, an individual with a low or high total outlet who feels fine is best regarded at the moment simply as an individual with a low or high total outlet—no more and no less.

Discrepancies in total outlet are not restricted to Western cultures. Ford and Beach reported that males of the Keraki (of New Guinea) generally had coitus once a week; of the Lesu (of New Ireland), once or twice a week; of the Hopi Indians, three or four times a week; of the Crow, every night; of the Aranda (of Australia), three to five times a night; and of the Chagga (of Tanganyika), up to ten times a night. Perhaps I should stop here before I lose my credibility.

Varieties of Sexual Experience

Like the frequency of total sexual outlet, its component behaviors are characterized by wide variability, significant correlations with biosocial variables (such as age, sex, or marital state), and the lack of causal explanations beyond the plausible assumption that biological, psychological, and cultural forces are jointly at play. Since we have used age differences as the illustrative example in discussing frequency of outlet, let us now examine the influence of two additional factors, marital state and educational level, as they apply to the six component behaviors contributing to total sexual outlet in Kinsey's findings.

Among single (never married) females, masturbation was found to be the single most important contributor to total outlet. In postpubescent girls below the age of fifteen, the youngest group, it accounted for almost 85 percent of all orgasms. By age twenty masturbation still provided about 60 percent of the total outlet. In the thirty-six to forty age bracket, masturbation had diminished to provide less than 40 percent, but by age fifty it was again providing over half of all orgasms among women who had never married.

Sexual intercourse was the next major source of orgasms for the single woman. Beginning with a low of 6 percent in the postpubescent group, it reached its peak between the ages of forty-one and forty-five, when it accounted for over 40 percent of all orgasms. Petting was responsible for about one in five orgasms between the ages of sixteen and twenty-five and continued to yield fewer but a still-significant ratio into middle age. Homosexual contacts provided a progressively larger proportion of orgasms until ages thirty-six to forty, when they accounted for a fifth of the total sexual outlet, then decreased again in subsequent age groups. Sex dreams were a minor source and animal contacts a negligible source of orgasms for this group.

The pattern of sexual behavior of married women differed dramatically from the above. At all ages sexual intercourse was decidedly the major outlet; some of it, however, was extramarital (highest in the forty-one to fifty age group where it accounted for over 10 percent of the total outlet). Masturbation, the only other significant outlet for this group, provided somewhat less than 10 percent of all orgasms across age groups. Other activities made small or negligible contributions to the total outlet for married women. The pattern for those previously married (separated, divorced, or widowed) occupied an intermediary position but was closer to that of the married group.

Comparison between males with different educational backgrounds demonstrated that education is a significant factor. Among unmarried men with no further education beyond elementary school, sexual intercourse was the leading contributor to the total outlet. In this group coitus had accounted for almost 40 percent of orgasms even among postpubescent boys who were no older than fifteen. Masturbation, the next major outlet, had provided about half of all orgasms after age twenty-one. Orgasm through homosexual contacts made up 10 percent of the total outlet up to age twenty-five, after which it increased to provide about a third of the orgasms between the ages of thirty-one and thirty-five.

In contrast, masturbation was the primary source of orgasms for single males with some college education. During adolescence it had provided fully 80 percent of the total sexual outlet. Although this ratio steadily declined, masturbation still accounted for almost half of all orgasms between the ages of thirty-one and thirty-five. There was far less reliance on coitus (especially with prostitutes) in the college-educated group. Up to age twenty, even sex dreams were a greater source of orgasms (about 10 percent) than coitus. By the ages twenty-six to thirty intercourse had moved ahead but still provided far fewer orgasms than other outlets combined. Petting accounted for a modest but steady ratio (about 5 percent) of orgasms. Homosexual contacts gradually increased with age, as in the less educated group, but at lower levels.

Among married men most of the above differences disappeared, and sexual intercourse emerged as by far the major source of orgasms for both college graduates and those whose education had stopped at the elementary school level. Whether operating through preselection or by virtue of the circumstances of married life, marital state obviously supercedes other factors such as educational level. Some slight differences nevertheless persisted between the two groups of married males; for instance, masturbation continued to be somewhat more prominent (about 10 percent of total outlet) in the more educated group.

Sexual Intercourse

Although there is much to be said *for* sexual intercourse, it is difficult to know what to say *about* it to people who have probably had the experience countless times. It seems hardly necessary to discuss the subject—yet books on this topic are selling in the millions. Most of this literature deals with two main topics: coital technique and the psychology of sexual intercourse. Although writers recognize the fact that these two issues are inseparable, it is standard practice to deal with them separately to facilitate discussion, and the discussion often gets stranded on the physical aspects of sex.

Coital Technique

Much of the advice of sex manuals about the techniques of sexual stimulation could be readily guessed by a few moments of reflection. We are told, for example, to start by stimulating relatively neutral areas and only then move to more sensitive and frankly sexual regions. Stimulation should be sustained in order to build up sexual excitement, yet a relentlessly repetitive persistence will create boredom and will actually lessen sensitivity in tissues. While one part is resting, another part should be stimulated so that some activity is going on most of the time. Ultimately the so-called erogenous zones are caressed each in its appropriate way: gentle touches for sensitive surfaces and firmer handling for large muscle masses. As excitement mounts, one can be bolder and more forceful.

Most physical stimulation involves the use of lips and hands. There are numerous forms of kisses and caresses, none of which seems to be the patented invention of any one culture. You can see Mochica pottery figures "French kissing," and there are more examples from various cultures of male hands fondling female breasts than anyone would care to count. Ancient Hindu manuals are rich in exotic procedures and specific directives about what to lick, nibble, bite, chew, caress, fondle, rub, pinch, tickle, scratch, bend, twist, press, push, and pull. You can also find out, if you must, how to coordinate all this with the phases of the moon.

Discussions of coital technique ultimately consider positions and movements of sexual intercourse. There is ample evidence that all cultures experiment with coital positions. Examples are found in cultures on all continents, in art that ranges from sacred temple carvings to less awesome depictions. The walls of Pompeiian houses of prostitution were decorated with paintings of copulatory sequences. These have been the inspiration for a famous set of postures by the sixteenth century painter

Giulio Romano and countless subsequent imitators. Even Leonardo da Vinci produced a drawing that shows the relationship of the sexual organs in a couple engaged in coitus. This art form is most elaborate in medieval Hindu temple reliefs showing gods and mortals in the most intricate sexual postures. In the Japanese *shunga* paintings and woodcuts (late seventeenth to early nineteenth century), depictions of sexual intercourse reach a level of art comparable to Greek erotic vase paintings in excellence. Lorenz Eitner reviews this historical record in his chapter on the erotic art in *Fundamentals of Human Sexuality.*

Apart from their artistic merit, what purpose do such illustrations and discussions of coital positions serve? Provided one does not become preoccupied with coital mechanics, they have considerable value. They add variety to sexual encounters that have a tendency to become routinized. "Marriage," noted Honoré de Balzac, "must continually vanquish a monster that devours everything: the monster of habit."

Although it would seem that the primary purpose of varying coital postures would be to elicit different physical sensations, there are equally important psychological advantages. Obviously, coitus in a face-to-face position brings into contact different surfaces of the body than would, for example, a rear-entry approach. In the former the couple see each other much better, and the female partner can be just as active as the male. Yet in the other alternative her breasts are more accessible to his hand. Similar significant differences exist among the other major choices whether the couple sits, stands, or lies down. But beyond a certain point such variation becomes redundant.

Apart from providing variety, experimentation with coital positions helps make intercourse a more deliberate and purposeful event. Routine sex is usually hurried as well as lackadaisical: to the bathroom, brush teeth, get into bed, a few perfunctory caresses, orgasm, lights out. (This sequence is described by a Southern California sex therapist as the "Wham, bam, thank you, ma'am" approach.) Let me hasten to add that there is nothing inherently wrong in such a sequence if a couple is happy with it. But most people seem to appreciate a somewhat more elaborate procedure at least once in a while, and experimentation with various positions makes this possible. Some of the more complex possibilities require pause, planning, and most important, communication and coordination—which is what successful lovemaking is all about.

Unlike coital positions, coital movements are usually underemphasized in the literature. You will recall that the two basic physiological processes underlying sexual excitement and orgasm are vasocongestion and increased muscular tensions. How one moves is thus more significant than how one is positioned relative to his partner. As we shall see further on, one reason some women fail to reach orgasm is that they do not bring their large muscle masses into action.

Pictorial representations of coital postures are in this sense misleading because they show couples locked in apparently frozen postures. Lovemaking is a fluid process involving almost constant movement of the whole body. In one continuous process various forces of stimulation overlap; one coital sequence leads to another until mounting tensions escalate to climax.

A number of findings from the Masters and Johnson research deserve mention because of their practical significance. One is the observation that the anatomical relationship of the clitoris to its overhanging cover makes it possible for the clitoris to be stimulated without direct contact. As described earlier, the minor lips split into two layers at their upper end and surround the clitoris. It is this upper layer, the prepuce of the clitoris, that rubs over the clitoris as the minor lips move to and fro during the penile thrusting that occurs in all coital positions. Until this fact was known, sex manuals devoted much effort to endorsing positions that would advantageously cause direct clitoral friction by the penis—difficult at best and, by what we now know, quite unnecessary. Additional clitoral stimulation, if desired, can best be done manually.

Masters and Johnson have observed also that, after the clitoris swells up during the excitement phase, it retracts under its hood in the plateau phase. Men who realize the erotic potential of the clitoris locate it early in lovemaking and guard it diligently. When the clitoris slips away in the plateau phase, they may erroneously conclude that the woman has "lost her erection" and is not excited any more, which of course is far from the truth.

The Psychology of Sexual Intercourse

The psychology of sexual intercourse has been the subject of considerable discussion, the essence of which is that there is more to sex than coitus and more to coitus than orgasm—as anyone with any sense already knows.

Irrespective of the outcome of current theories linking sexual behavior to instinct or hormones, it is inconceivable that we will ever understand sexuality without understanding its biological roots. This does not mean that sex has a "purely" physiological basis and that all we need to do is to discover it. Even eating, so obviously physiologically motivated, is subject to psychological and cultural forces that sometimes override biological needs. It is thus misleading to speak of biological determinants as if they were relentless forces automatically "driving" and manipulating people like hapless puppets.

While we have to live with the mystery of biological givens for the moment, a number of psychological aspects of sexual behavior are worth our attention. First and foremost is pleasure. Animals engage in coitus because it is pleasurable, and so do we. There is no rational way that we can romanticize, philosophize, or theologize ourselves out of this evolutionary heritage. But whereas lower animals may stop there, human expectations of coitus go far beyond carnal pleasure or relief of tension. Sexual intercourse, like any other form of human interaction, is inevitably colored by the characteristics of the persons involved: universal or prevalent human qualities modified by the idiosyncratic characteristics that mark the individual.

Most people find it necessary to have some degree of affection for the sexual partner if the encounter is to be rewarding beyond mere orgasmic release. Beyond that, opinions vary. Some sincerely believe (and many others profess to believe) that sex without love is demeaning. Others do not find such profound affection any more essential than it would be for a pleasant evening over dinner or a game of tennis with a congenial partner. The matter goes beyond the issue of pleasure and becomes a standard of morality. A fairly common attitude, especially among the young, is that "it is all right if you are in love."

People have written about many kinds of love, erotic love being only one. But even erotic love has a number of components in addition to sexual gratification, such as the fulfillment of dependency needs and the enhancement of self-esteem. The roots of our dependency needs —the need to be cared for—go back to our early infancy when we were totally dependent on our primary caretakers for our physical and psychological survival. Although most of us manage to survive the loss of this earthly paradise, these same early needs continue to crave satisfaction and part of it is achieved through sex. Women are more open in wishing to be held, hugged, and cuddled. Some women will admit that they engage in sex for the gratification of such needs while other aspects of sex leave them cold.

Despite the cool and self-reliant façade that men are expected to maintain in our culture, they share these dependency needs. It takes a perceptive woman to respond to these needs with the necessary tact and subtlety, just as it takes a sensitive mother to know how and when to hug her tough little boy without "sissifying" him. (I use these terms in the context of our current conception of sex roles and stereotypes.) In the light of these considerations, the endearments and caresses of foreplay take on additional significance. This is why there is need for approaching these activities with something more than the "erogenous zones" mentality.

In the postorgasmic phase there is a resurgence of these longings. One can think of no other time when grown men and women act more pacified, more like newly fed infants, than they do following orgasm. Yet this state of contentment and joy (when women are said to be at their loveliest) is sometimes tinged with sadness. Some people actually cry. There are cases where this feeling is due to remorse ("I did it again!"), regret that one is not in bed with someone else, recrimination over having been exploited, and other similar feelings. But more often the person has no conscious reason to feel sad or subdued. Perhaps this feeling is another manifestation of the "crying with joy" phenomenon and indicates an expression of diffuse and mixed emotions.

Like dependency needs, self-esteem is nurtured through a variety of rewarding experiences of which sex is only one, albeit an important one. Although self-esteem is a basic human need, its enhancement is subject to a host of cultural and personal circumstances. Impotence, one of the more humiliating experiences in a man's life, may simply bruise one person's pride while devastating another to the core. Lack of orgasmic response presumably enhanced a Victorian lady's self-esteem, but it is distressing for most modern women.

In successfully taking over control of society, males have accrued a number of liabilities, one of which is this close association of self-esteem to sexual competence. Although cultures differ in their preoccupation with *machismo*, this concern is present in even the most emotionally glacial circles. A man may fail elsewhere, but he is expected to be able to do the job in bed. So susceptible are men in this regard that most cases of impotence before old age are due to psychological causes. It is not uncommon for a man with problems of potency to feel incompetent in areas that are totally unrelated. Occasionally a person in this situation may unconsciously overreact with an overwhelming desire to be all-powerful in some compensatory activity.

The man is more vulnerable because the mechanics of the sex act place the primary burden on him. This vulnerability can be dealt with kindly—this ensuring the man's lasting gratitude—or aggressively. The most effective way for a woman to enhance a man's self-esteem is by being sexually responsive to him. While one can argue the ethics of faking orgasm, it is hard to make a case against giving a man all possible benefits of the doubt in cases of less than distinguished performance.

While I have no formal studies to cite in support of this assertion, it stands to reason that a woman would tend to be vindictive of male failure in inverse proportion to the tenderness and care lavished upon her during the preliminaries. A healthy woman will want emotional and physical gratification from a man (and vice versa), but if worse came to worse, one wonders if most women would not prefer a compassionate man with failing powers to a stud with four left feet.

Love and hate may not be exactly two sides of the same coin, but sex is a currency common to both. Not only do people make love as well as war; they also wage battles while making love. Resentment, anger hatred can be expressed during sexual intercourse, and such feelings heavily color the experience. To document this, examine our language. The most common English vernacularism for sexual intercourse also means to be taken advantage of. In various forms, the word also has other distasteful meanings—unpleasant, inferior, difficult, confusing, blundering, or "neurotic."

There are countless ways to punish another or oneself through sex. Some means are passive: the person is "too tired" or just doesn't feel like it—even when there are no legitimate grounds. Perhaps a couple returns home from a party where the husband drank too much or made passes at other women; now he fumbles for his wife, who feels aroused but refuses to play and may or may not give her true reasons. In another instance, the wife may have made some unreasonable demands or a cutting remark over dinner. At bedtime he approaches her more or less as usual, but there are subtle differences: he cuts foreplay short or is perfunctory about it; during coitus he may be more eager to get on top. He may act a little rougher or fall asleep faster after orgasm. All this may or may not escape her notice, and he himself may be hazy about the significance of his behavior. Depending upon the larger context of their sexual lives and overall relationship, such events may be soon forgotten, or the cause of long faces at breakfast, or the starting point of horrendous rows.

People with more serious problems go through life suffering a variety of malfunctions or perversions of the sexual impulse: women systematically undermining men through sexual enticement; men misusing, exploiting, or assaulting women. This is the stuff of which a good deal of the world literature from its finest to its lowest is made, and it reflects a very real aspect of life. The psychiatric literature also is suffused with such material viewed from a clinical perspective.

So far the focus has been mostly upon the ordinary, everyday aspects of sexual intercourse, what most of us experience most of the time. Now we shall briefly turn to two additional aspects: First, the enhancement of sexual pleasure, the "gourmet cooking" of sex, if you wish. Second, sexual malfunction, the problems that occur when sex does not work as well as we are generally entitled to expect (when the cake does not rise—to persist with the dubious culinary metaphor).

Enhancement of Sexual Pleasure

Sane and sensible men and women take sex in their stride as a natural right to be enjoyed under legitimate circumstances. Currently this premise is being extended to the assumption that a person owes it to himself to get the maximum charge out of sex if he is to live out his full "human potential." While this view may be an admirable antidote to residual Victorian shackles, it may also impose its own burdens on people. Whereas in the past those who sought treatment for sexual problems usually suffered from lack of orgasmic response or potency problems, a new kind of patient seems to be emerging: one with an apparently functional sex life who nevertheless is dissatisfied with it.

There is nothing inherently wrong, I suppose, in setting one's sights ever higher, provided this does not become an obsession. It is one thing, however, to fantasy about exploring the jungle and another to let it spoil your walk on the beach. While it is not my intention to endorse mediocrity, common sense dictates that there are trips other than safaris that are also worthwhile. Those bent upon exploring the wilder shores of love need to remember that a relentless search for the "perfect orgasm" can be as futile as the search for perpetual motion machines, fountains of youth, and assorted Utopias.

There is a great deal of loose talk these days about what sex can do for you, but my reading of the literature has failed to convince me that "superior" sex will make you a "better" person. Sex is a physiological function with profound psychological meaning, but it is no religion. It offers pleasure and enrichment but no salvation. If you are going to do it, you might just as well do it right—but this is not necessarily going to advance the cause of mankind.

Assuming one is setting out with sensible goals and fortunate enough to have a spouse who shares one's interest and drive in this regard, what is the available advice about enhancement of sexual enjoyment? I hate to disappoint you, but as far as I can ascertain, the state of the art does not amount to very much. But I must hasten to add it is by no means insignificant and therefore worth some attention.

Let us consider, for example, the advice of two recent books: *The Sensuous Woman* by "J" (Joan Garrity), which appeared in December 1969, and *The Sensuous Man* (1971) by "M," who has remained anonymous. Unlike the love manuals that were addressed mainly to princes with harems and not much else to do, these current counterparts speak to ordinary folk who are assumed to have humdrum sex lives.

Both "J" and "M" claim to have become sensuous from rather inauspicious beginnings. "J" writes that she has ". . . heavy thighs, lumpy hips, protruding teeth, a ski-jump nose . . ." and is shy. Yet she says, "Some of the most interesting men in America have fallen in love with me," and adds that she has learned to be sexy, all woman, ". . . a lady in the living room and a marvelous bitch in bed, sensual, beautiful, a modern Aphrodite, maddeningly exciting, the epitome of the sensuous woman." "M" in turn was twenty-eight years old before he "really" learned how to make love, a waste of thirteen years by his reckoning. The message is, "You can do it too."

The combined advice of the sensuous authors is part general and part specific. One is told to shape up: exercise, keep clean, do not smell bad, do not "sound like a fingernail screeching across a blackboard," do not spray saliva when you talk, nor allow foam to gather in the corners of your mouth. The more specifically sexual advice starts appropriately enough with the premise that sex is "all in your head" and goes on to point out the need to be considerate, sensitive, and unabashedly frank and joyful about sex.

Then come "sensuality exercises" aimed at heightening sensitivity. Sample advice to women: Following a hot bath, stretch on the bed naked and lovingly massage yourself with a lotion. Other exercises are designed to strengthen the muscles of the pelvis or the tongue. Sample: Stick your tongue out and move it left to right like a windshield wiper while touching the edge of the mouth.

A notable divergence from the more traditional marriage manuals is an open endorsement of masturbation that borders upon the exuberant. Masturbation is "sensuality exercise" No. 10 for women and, because of its importance, it rates a special chapter. Masturbation is recommended as fun and as excellent training for coitus. Detailed instructions range from manual techniques to the use of vibrators and the Jacuzzi whirlpool bath.

In addition to the customary advice on foreplay, these books give specific instructions for mouth-genital stimulation for both sexes. As early as 1926, the Dutch gynecologist Van de Velde, author of *Ideal Marriage*, boldly endorsed the "genital kiss" but said little about how to do it. His modern counterparts describe it with a vengeance. The male tongue is instructed in the intricacies of the "alternate flame," the "feathery flick," and the "velvet buzz saw." The sensuous woman is told how to stimulate the male orally by the counterparts of the above: the "butterfly flick," the "silken swirl," and the "hoover" (which is exactly what you think it would be). To top it off there is the "whipped cream wriggle."

Coital positions are treated in fairly standard fashion, since all that can be said about these has been said before. There are no claims of "new" positions nor the implication that there are one or two positions to end all positions. Enhancement of the coital experience is further discussed in terms of when and where to make love.

Then there is the whole area of erotic aids, including soft lights, alluring garments, music, perfume, champagne, and everything else to which we commonly ascribe erotic powers. The use of lotions has recently been recommended as an effective enhancer of foreplay.

The more venturesome try vibrating beds and, more recently, water beds. The use of mirrors is centuries old. There are condoms fringed with projections, rings with prongs that fit over the glans, and assorted other devices to tickle, prick, scratch, and variously stimulate the vagina during coitus. Although most people make do with household pillows to assist them in lovemaking, the more determined may have stools and other special furniture around to execute the more complex coital variations. I have seen the photograph of a finely constructed and upholstered contraption built like a double-decker gynecological examination table, which apparently was used by an amorous English aristocrat.

Vibrators are recommended as highly effective erotic stimulants. Some fit on the hand and transmit their vibrations through the fingers. Others are more like electric toothbrushes minus the brushes or with rubber tips of various shapes. These instruments are used in foreplay as well as to achieve orgasm.

Modern sex manuals do not discuss the use of pain as an erotic stimulant, probably because of touchy associations with sadomasochistic practice. There is one interesting example of the use of cold (a "celebration special") suggested in *The Marriage Art* by J.E. Eichenlaub: As a person is going into orgasm, his partner picks up a cold towel or a handful of crushed ice and jams it at the genital area. The impact is said to be remarkable.

Finally a few odds and ends:

One often hears about aphrodisiacs, substances that presumably enhance the sexual drive and performance. While the possibility that such substances exist cannot be dismissed out of hand, none has been convincingly demonstrated. Excitement derived from sights, sounds, smells, and tastes is a learned response and does not qualify the stimulus as an aphrodisiac in a strict sense. By relaxing the individual and removing inhibition, alcohol in small quantities helps some to act in a more outgoing and lively fashion sexually. The same is true of marijuana. Small doses of mild tranquilizers may calm an anxious person and thus help his sexual functioning. Used in excess—and in some cases even in moderate doses—all of these substances will depress sexuality.

If most of the advice on erotic enhancement consists of the commonplace, and the rest borders on the exotic if not the idiotic, who needs it? Paradoxically, many people seem to need it, and I think such advice fulfills a number of functions. First, it seems to satisfy a general curiosity, naïve as it may be. When I read one of these sex books, I realize how interested I would have been as a teen-ager if I had discovered one then; since many of us never did, perhaps we are perpetually trying to fill that void. Another motivation in reading such advice is that it is in itself potentially arousing.

Finally there is the power of the printed word. A few moments' reflection would probably enable us to figure out most of what these sources reveal. But we do not generally do that; perhaps we do not want to see ourselves as sexually brash. When a procedure is sanctioned in print, one is not entirely responsible for it and feels freer to adopt it. Or merely reading about it may provide some vicarious gratification.

Sexual Malfunction

If the vagaries of enhancing sexual pleasure frequently convey the whimsy of a pseudo-problem, chronic sexual malfunction is no joke. There is probably no other area of life that entails more silent suffering. Many with problems of this nature are embarrassed to seek help, and those who work up enough courage may not know where to turn.

Kindly, broad-minded general medical practitioners and clergy are often the first to be consulted, but most of them must rely on personal experience, common sense, and fragments of knowledge to deal with these problems. Even gynecologists, urologists, psychiatrists, and marriage counselors are often less confident of their skills in this area than in the rest of their practice. A handful of "sexologists" concentrate on these problems, but as a group they are hardly the equal of other specialists in prestige and influence. I do not know of a single medical school that has a department or even a full-fledged professor of human sexuality. This situation is changing, however, and it is probably only a matter of time until sex specialists occupy their rightful place.

There is an extensive literature on sexual malfunction in psychoanalytic and psychiatric writings, in sex manuals, and in publications by (or influenced by) Masters and Johnson. The general literature also is full of stories of men and women with disturbed sex lives.

The term "sexual malfunction" is usually restricted to disturbances of sexual intercourse itself, rather than variant or deviant patterns of sexual behavior such as homosexuality. Yet coital difficulties are not discrete disease entities. Rather, they are symptom complexes of varying severity caused by physical (rarely), psychological (most often), and cultural (difficult to substantiate) factors. Although it is customary to discuss disorders of males and females separately, it is preferable to view coital difficulties as the joint problem of a couple.

Sexual apathy is one form of malfunction. Obviously no one is expected to engage in sex any time, any place, and with anyone; it is thus perfectly normal and quite common for a person to feel occasionally disinclined when tired, worried, sick, or preoccupied—or even for no apparent reason. But if one is chronically disinterested even under propitious circumstances when a legitimate and interested partner is available, then we wonder why. If the person's feelings go beyond mere indifference and are tinged with anxiety and dread, something is usually wrong.

More commonly, sexual malfunction implies coital inadequacy, not just lack of interest. Among males this usually takes the form of impotence (inability to attain or maintain an erection) or "premature ejaculation," a rather vague concept whereby a man cannot delay orgasm "long enough." Pain during coitus, rarely a problem for males, is a common complaint of women with coital inadequacy (and is called *dyspareunia*). Women have no problem with "premature" orgasm, since they have no erection to lose, but they are sometimes troubled with the inability to reach orgasm (frigidity). As indicated earlier, the differentiation between so-called vaginal and clitoral orgasms is physiologically meaningless; furthermore, many women seem to be readily multiorgasmic, although a single orgasm also may be perfectly satisfying. Frigidity then must be taken to mean failure to reach orgasm of any kind.

There are in addition some rarer disturbances. The vaginal introitus may become so spastic that penetration proves impossible (*vaginismus*). Or a male may ejaculate in "retrograde" fashion whereby his semen flows into his urinary bladder rather than being discharged to the outside. Another rarity is persistent and painful erection, often without sexual desire (*priapism*). Or a man with a normal erection may be unable to ejaculate.

Impotence and frigidity are further divided into primary and secondary types, although the differentiating criteria may vary. Usually the term "primary" is reserved for cases in which the person never attains erection or orgasm, whereas in cases of "secondary" dysfunction, erection or orgasm may be reached in some activities or on some occasions. Or these terms may be used exclusively in conjunction with coitus and no other activity.

Like sexual desire, male and female sexual response waxes and wanes and hardly anyone escapes occasional inability to produce an erection or failure to climax. Because of this, it is difficult to set specific limits at which function would be regarded as disturbed. Masters and Johnson consider a man secondarily impotent if he regularly fails at coitus in one out of four instances. Even more arbitrary is the definition of "premature ejaculation." Obviously a man who usually ejaculates immediately after penetration and against his own will is "premature." But beyond that, how long should he last? Kinsey reported that three out of four men reach orgasm within two minutes after intromission. Whether any of them would be considered "premature" depends to a large extent upon the reaction of his partner. If the woman has reached orgasm by the time of intromission, she may be quite satisfied with his performance (or even glad to get it over with). If his orgasm occurs before she has had a chance to become aroused, however, she is likely to be quite dissatisfied—hence the need to evaluate the couple's interaction rather than to stopwatch individual performance.

In view of these ambiguities, it is difficult to obtain rates of potency problems in the population. An additional factor is the fact that males gradually lose their potency with age, although such decline is not inevitable and there is much variation. Within these limitations it has been noted that one out of every hundred males below the age of thirty-five is impotent; by seventy years of age, one out of four is impotent.

The problem of achieving erection may coexist with the inability to control ejaculation. Men with such ineffectual performances may forego attempts at coitus to avoid humiliation. Here is a description of a man "contemptible in the eyes of women" from the eleventh-century book *The Perfumed Garden* by the Moslem sage Shaykh Nefzawi:

> When such a man has a bout with a woman, he does not do his business with vigour and in a manner to give her enjoyment. He lays himself down upon her without previous toying, he does not kiss her, nor twine himself round her; he does not bite her, nor suck her lips, nor tickle her.
>
> He gets upon her before she has begun to long for pleasure, and then he introduces with infinite trouble a member soft and nerveless. Scarcely has he commenced when he is already done for; he makes one or two movements; and then sinks upon the woman's breast to spend his sperm; and that is the most he can do. This done he withdraws his affair, and makes all haste to get down from her. . . . Qualities like these are no recommendation with women.

In the past, lack of orgasmic response among women presumably caused little concern. This was in part because neither the mechanics of the sex act nor fertility is affected by lack of female orgasm. Furthermore, some nonorgasmic women seem to enjoy sex. It is also pertinent that whereas male orgasm has a long evolutionary history, among female animals, even nonhuman primates, evidence for orgasm is scanty and often ambiguous. In the male ejaculation proves that orgasm has occurred; in the female there is no similar evidence. But currently attitudes are changing drastically, with women expected to be equal recipients of the pleasures of coitus.

That sexual problems are involved in a large proportion of troubled marriages is generally agreed upon by workers in the field. It is not always easy to show whether the sexual difficulty is the cause or the effect of the poor relationship. There are probably bad marriages in which sex is no problem as well as good marriages in which sex leaves something to be desired. Gebhard has reported that in marriages considered "extremely happy," almost 5 percent of the wives were nonorgasmic with their husbands. In general, however, no one would deny that a good sexual relationship tremendously strengthens a marriage, and a poor sexual adjustment seriously stresses it.

The causes of sexual malfunction are often difficult to unravel. In a relatively small number of problems the cause is physical. This is especially true if there is pain, and any woman who experiences pain during coitus needs to have a careful gynecological examination. It is important to differentiate pain on intromission, which tends to be sharp, from a duller pain that occurs during and following coitus. Physical problems in male dysfunction are rarer, but again a careful inquiry and examination are in order.

Most sexual dysfunction in both sexes has no demonstrable physical basis and is ascribed to psychological causes. Anxiety and depression, both inimical to sex, are examples of *intrapsychic* problems (caused by internal conflict). The problem is sometimes *interpersonal* (conflict between the sexual partners) and therefore more likely to interfere with one relationship than with another. Obviously these two aspects are often inseparable.

The specific psychological problems singled out depend upon the therapist's theoretical orientation. Behaviorists ascribe malfunctions to faulty learning, while psychoanalysts blame a long-repressed conflict. In many instances problems are traced to the Oedipal situation; if a person retains an unconscious erotic attachment to a parent, he will react to sexual partners as if they were incestuous objects. In this situation one may shun sex (by becoming impotent) or be unable to enjoy it (frigidity). The impact of cultural influences upon sexual function is a fascinating but problematic issue, which we shall consider in the next chapter of this book.

The treatment of sexual malfunction depends on its demonstrated or suspected cause and the therapist's professional background. Physical problems are managed by medical methods. Symptoms attributed to psychological causes may be treated by behavior therapy aimed at correcting faulty learning; or they may be dealt with by methods of psychotherapy ranging from brief counseling sessions to full-scale psychoanalysis. The most significant recent advances in this field have come from the research of Masters and Johnson. Since concise and clear expositions of their work are readily available, I shall refer you to these sources rather than summarizing them here.

A simple procedure discovered almost accidentally by A.H. Kegel, a gynecologist, has reportedly helped women to overcome difficulties in attaining orgasm. It consists of a set of exercises involving the muscles of the vaginal introitus. To learn where and how these muscles function, the woman is instructed to stop the flow of urine voluntarily a few times. The exercise itself, repeated five or six times a day, consists of flexing these same muscles rhythmically, beginning with sets of ten. Kegel had been using these exercises to help women regain control over the urinary incontinence that sometimes follows pregnancy. After his patients reported spontaneously that their sexual functions seemed to be improving, Kegel started recommending these exercises for cases of orgasmic unresponsiveness.

Variations and Deviations of Sexual Behavior

In judging any form of behavior, the questions most often asked are: How common is such behavior? Is it healthy? Is it moral? Is it legal? The answers tend to be closely related, and judgments are often interdependent. When applied to sex, answers to these questions tend to be uncertain, relativistic, and more often sustained by feeling than substantiated by fact. There are also wide discrepancies between what we say and what we do, and what we permit some but not others to get away with.

It is thus very difficult to delimit what is "normal," what is an acceptable "variation" of the normal, and what is an unacceptable "deviation." For the purpose of this chapter, I propose to omit the ethical and legal aspects and deal only with the medical or health perspective, along with further consideration of the issue of prevalence.

In 1905 Freud published a short work entitled *Three Essays on the Theory of Sexuality*, which attracted little notice at the time—and earned Freud the equivalent of $53.08 in royalties over a ten-year period. These essays have come to be regarded as one of the most significant contributions to psychoanalysis. It was in these essays that Freud outlined his theory of infantile sexuality and his model for all subsequent forms of sexual behavior. The evidence appeared so convincing to him that Freud wrote in his introduction: "If mankind had been able to learn from a direct observation of children, these three essays could have remained unwritten."

Psychoanalytic views of human sexuality see the infant as being endowed with a biologically determined sexual instinct or drive. This sexual force, or libido, is successively invested in the oral, anal, and the genital regions of the body. Eating and elimination, in addition to their life-sustaining functions, provide the child with pleasure that, like all pleasure, is fundamentally "sexual," the term being used in a far wider sense than is generally understood. At about the age of three, the genitals are the erotic focus. Because it is specifically the penis and the clitoris that are so invested, this stage is known as the phallic phase of psychosexual development.

Since oral and anal functions are similar in the two sexes, "pregenital" development proceeds the same way for both; but with the phallic phase, male and female developmental paths depart. This phase is complicated by the emergence of the famous Oedipus conflict whereby the child is unconsciously attracted to the parent of the opposite sex in competition with the parent of the same sex. Normally this conflict is resolved and the residues are buried in the unconscious. The child then proceeds through a period of latency or relative sexual quiescence while gradually developing "genital maturity" when his attention shifts from a narcissistic interest in his own organs to the genitalia and ultimately the person of an adult member of the opposite sex. The healthy person thus longs for heterosexual intercourse as the means of gratifying his sexual needs with his chosen mate or sexual "object."

Until this adult solution is achieved, the child and to some extent the adolescent is free to obtain sexual gratification through all sorts of means and objects—children are thus normally "polymorphously perverse." (This "defamation" of childhood innocence did not and does not endear Freud to many.) Large numbers of people never quite make it to the adult norm but cling to one or another of these infantile modes of obtaining gratification. If they do this in a distorted, camouflaged manner, the result is a "neurosis"; if the infantile sexual wish is permitted more or less free and undisguised expression, the result is a "perversion."

These perversions or deviations of the sexual impulse, which are nothing more than persistent patterns of infantile sexuality, may take one of two major forms: deviations in the choice of sexual object or deviations in the sexual aim. Under the first category the normal object (an adult of the opposite sex) could be replaced by an adult of the same sex (homosexuality), a child (pedophilia), a close relative (incest), an animal (zoophilia), an inanimate object (fetishism), or even a dead person (necrophilia). Under deviations of the sexual aim, the person would view (voyeurism), expose himself (exhibitionism), hurt (sadism), or suffer pain (masochism) rather than have coitus with the sexual object. When both object and means are deviant, the behavior is classified according to the choice of object.

This is a neat way of gathering everything under a single conceptual umbrella, but it has shortcomings, including questions about Freud's fundamental assumptions. Furthermore, this theoretical scheme must be reconciled with practical considerations. Obviously, a person who would rather have orgasm through mouth-genital contact is very different from a sex murderer, and masturbation against a pillow is not the same as copulating with a goat, even though all four alternatives deviate from the psychoanalytic adult sexual norm.

For all practical purposes, therefore, what we label as deviant is to a large extent determined by the social consequences of the act rather than by theoretical considerations. Another important qualification is that these "deviations" are exceedingly common as mild, scarcely recognized tendencies; for example, few men would not steal a glance at an upturned skirt, and some element of pain is often enjoyable in coitus. It is only when these alternatives become a necessary condition or an exclusive substitute for coitus that we mark them as deviant—with considerable arbitrariness at that.

Some of the more common accompaniments and alternatives to coitus are as follows:

Erotic fantasy is probably the one sexual activity that is truly universal. The prescription of medieval penances for the offense (twenty-five days for a deacon, fifty for a bishop) indicates that even the celibate and chaste clergy are not considered exempt from the temptation. In sexual dreams, another form of fantasy, latent wishes appear in variously distorted, symbolic forms. The so-called nocturnal emissions ("wet dreams") are sexual dreams that end with orgasm. Although women do not ejaculate, they too have such dreams.

Nocturnal orgasm (any orgasm occurring during sleep) accounted for only 2 to 3 percent of the female and 2 to 8 percent of the male total outlet in the Kinsey sample. About 5 percent of men and 1 percent of women had weekly orgasms while asleep. This evidence does not support the contention that if one would refrain from all sex, the body would use this "safety valve" to release sexual tension.

A fascinating new field has been opened recently through studies of the neurophysiology of sleep and dreaming. It has been shown that we all dream every night whether or not we recall our dreams in the morning. It has been observed also that in nine out of ten cases males show erections during these active sleep periods, which are identified as REM periods because they are marked by rapid eye movements. (One of the pioneers and leaders in this field of research is William C. Dement at Stanford.)

Masturbation is another widely (but not universally) practiced sexual outlet. It has come to be accepted "officially" as harmless and normal even though some doubts linger on in both lay and professional minds. Perhaps more common is the view that there is something funny and embarrassing about it: Philip Roth's *Portnoy's Complaint* is a case in point. Historical references to masturbation go back to the Babylonians, Egyptians, Hebrews, Indians, and Greco-Romans. Even Zeus was said to indulge occasionally. The more recent history of Western attitudes toward masturbation, a remarkable story, has been described in detail: H.E. Hare, "Masturbation Insanity: The History of an Idea," *Journal of Mental Science*, 452 (1962), 2-25. See also Chapter 3 in: Alex Comfort, *The Anxiety Makers*. New York: Dell Publishing Company, 1970.

Here is a brief review of the statistics of masturbation: Handling of the genitals starts in infancy, but deliberate self-stimulation to orgasm usually begins in the early teens. Ultimately 95 percent of males and almost 60 percent of females reach at least one orgasm through it. By age eighteen or so most of the males who will ever masturbate have already done so, but in females there are new recruits all the way to age forty-five. The practice is more common among the better educated; it is unrelated to religion among males but less common among devout females; women born after 1920 are more likely to masturbate than those born earlier, and more urban women than rural women adopt this outlet.

Masturbation is not a poor man's coitus either literally or figuratively. It is reported from many cultures whose people have free access to intercourse, although it is disapproved for adults in many of them. Self-stimulation has been observed in many animals including porcupines, dogs, cats, elephants, and dolphins, as well as primates; generally males excel at it.

Although manual techniques are most common for both sexes, friction against objects is frequently used. Women may reach orgasm without touching their genitals through rhythmic pressure of their thighs, and vaginal insertions are used by about one in five. Fingers are most often relied upon, but any conveniently shaped object can be drafted into service. Artificial penises or "dildos" are more often used for male entertainment than for private pleasure. Modern versions come battery powered. Males can avail themselves of vinyl "masturbatory dolls" fully equipped with various options such as pubic fur. As masturbation becomes more accepted, I suppose we can look forward to additional assistance from our highly advanced technology.

Most of the sexual practices more commonly thought of as deviations are rare and of relatively little concern to the nonspecialist. An important exception is homosexuality, an area in which important attitudinal changes are taking place. Homosexuals traditionally have been viewed as sick. While this viewpoint is still dominant, a trend that is gaining ground is to view homosexuality as an alternative life-style. This view is supported by the lack of convincing evidence to date of any hormonal or physiological disorder in homosexuals and the contention that many homosexuals would be quite satisfied with their lives if they were left alone.

While homosexuals continue to be hunted and hounded in many places, in more liberal urban enclaves some homosexuals now lead lives that are fairly free, and a few enjoy national prominence without having to conceal their sexual inclinations. In Britain, among other countries, homosexual contact in private between consenting adults is not a legal offense—which is not to say that the British think homosexuality is healthy or moral.

Much of what is generally "known" about homosexuality is now being shown to be inaccurate and the result of years of distortion and misconception as well as the direct outcome of persecution and harassment. As a result one can at present hardly generalize without a high risk of error.

Kinsey staggered his contemporaries with his statistics on homosexuality: 37 percent of males and 13 percent of females in his sample had experienced at least one homosexual orgasm; among the single, the corresponding figures were 50 and 26 percent, and among married persons, 10 and 3 percent. Kinsey devised a heterosexual-homosexual rating scale showing the ratio of orgasms attained by these respective means and felt that it conveyed a more accurate picture of homosexual behavior in the population by illustrating the gradations that exist between exclusively heterosexual and exclusively homosexual behavior.

(This scale is discussed briefly in Chapter 11 of *Fundamentals of Human Sexuality* and in much greater detail in the chapters on homosexual behavior in the Kinsey volumes.)

Kinsey's statistics represent a lot of homosexuality—and people still react with incredulity to these figures. Given the shortcomings of the Kinsey survey, no one can make absolute claims for the veracity of these data, but findings from other studies also turn out rates high enough to make people feel uncomfortable. About 1 percent of inductees are rejected by the armed forces because of homosexuality. While some may be faking in order to avoid military service, a significant number also gets into the armed forces by concealing homosexuality. By the most conservative estimates, homosexual orientations can be attributed to several million American males; as a rule of thumb, figures for females tend to be about a third of those for males.

Who are these people? A composite picture emerges from the scanty authoritative literature, educated guessing, and the nonproselytizing homosexuals who are willing to talk. (We are concerned here with reasonably confirmed though not necessarily exclusive homosexuals, rather than with some guy who sometime, somehow, ended up briefly holding someone else's penis.) Homosexuals, like heterosexuals, may be short or tall, muscular or puny, stupid or smart, kind or cruel. There is no reliable physical or psychological trait that will sort them out. Because some homosexuals flamboyantly affect certain movements or postures, they reinforce the popular myth that there are specific, telltale homosexual mannerisms or distinguishing physical features. Like all persecuted groups, homosexuals develop subtle (and not so subtle) ways of recognizing each other, but these signals are learned. One should remember that there are men with high-pitched voices, limp wrists, and mincing gaits who are not homosexual.

The individual's life-style is influenced by whether he is an overt or covert homosexual and by a variety of other factors. A homosexual college student, for example, is more likely to depend on social contacts rather than frequenting gay bars. Although homosexual liaisons between men tend to be of short duration, longer associations exist, and homosexual jealousy can be intense. Serious sexual encounters usually consist of mouth-genital contact and anal intercourse, which may be preceded by foreplay. It is generally meaningless to assign "active" or "passive" roles to these acts. Incidentally, there is no such thing as a specifically homosexual act—what defines the relationship is the identical sex of the participants. Heterosexual couples also can engage in oral-genital contact or anal intercourse.

Lesbian liaisons tend to be more lasting and less likely to attract suspicion. Sexual activities usually involve petting, mutual masturbation, and mouth-genital contact. The use of dildos is rare—contrary to popular belief. It has been maintained that lesbians more consistently stay with specific roles ("butch" or "femme"), but apparently this is changing now in the direction of less differentiated relationships.

Homosexuality is not an all-or-none phenomenon. Some persons are "bisexual" and engage in relationships with both sexes; many are married and have children. Homosexuals should not be confused with transsexuals, who sincerely believe that, despite their anatomic gender, psychologically they are like members of the opposite sex. Transsexuals are usually male, and in recent years increasing numbers have undergone sex surgery. The combined effect of hormonal treatment and surgery sometimes has spectacular results. It also is possible now to make an anatomic male out of a female. (Donald R. Laub at Stanford Medical School is one of the recognized experts at sex-change surgery.)

Transvestites are not necessarily homosexual. Although some homosexuals dress in the clothes of the opposite sex ("queens in drag"), not all males who so dress for erotic or burlesque purposes express conscious homosexual wishes or engage in such behavior.

CHAPTER THREE

SEX AND SOCIETY

AN IMMORTAL fantasy assumes the existence of societies in which sex is "free"—or at least that there were such societies before "civilization" ruined them. In these blissful settings one could presumably have sex any time, any place, in any fashion, and with anyone, just as naturally as one would pluck a banana or clamber up a tree to feast on its fruit. The existence of such societies has never been documented; it appears that every society, in one fashion or another, attempts to regulate the sexual behavior of its members.

Although it is legitimate to compare various cultures in terms of levels of sexual permissiveness, such comparisons must be qualified by specifying the practices with which we are concerned. A primitive culture, for example, may be quite liberal in allowing sexual unions without benefit of clergy, but may then greatly restrict all sexual relations by expecting people to abstain during the harvest season, prior to a raid, or for so many days after the death of a chief.

In Chapter 10 of *Patterns of Sexual Behavior*, Ford and Beach examine cultures in terms of their regulation of sexual activity among the young. As a result of their survey, they classify fourteen cultures (including American society) as restrictive, forty-eight as semirestrictive, and thirty-two as permissive.

Even among nonhuman primates sex is neither random nor unrestricted. Access to sexual partners is regulated by factors such as the physiological functions that determine female receptivity and the dominance pattern of the troop.

Despite intercultural and individual differences, there are some remarkable similarities among societies. Incest, for example, is universally prohibited. Heterosexual intercourse is the dominant adult sexual mode, even though one can find all other forms of sexual expression in cultures past and present.

Although all societies have sexual laws, these seem to be applied with less than even-handed uniformity: what the aristocracy flaunts would send the pauper to the gallows; an artistically gifted individual may be forgiven his not-so-secret vice, but the same behavior in another may effectively end his career; it is one thing for a man to keep a mistress, and quite another for a woman to step out of line. Another universal is the remarkable discrepancy between what is said and what is done. Finally, where sexual permissiveness is concerned, the grass usually looks greener elsewhere. Americans think of Europe as lusty and decadent, while Europeans think of America as wild and uninhibited. The West imagines that the East is free of Victorian constraints; the East considers the West the contemporary Sodom and Gomorrah.

This network of contradictions is tangled by the factor of change. Each period of history makes its claim that "things are changing" and records the responses of dismay or hope that this claim elicits. Each era feels that its own throes of change are more real and more radical; for example, we are now supposedly going through no less than a "sexual revolution." How can one evaluate such claims? We are tempted to gain comfort from the ancient wisdom that "there is nothing new under the sun" and to invoke the truism that the more things appear to change, the more they in fact remain the same. There is much to be said for maintaining the calm that comes from an historical perspective—as long as it does not deteriorate into a head-in-the-sand stance. But society should be alert to changes and should even try to anticipate them in order to deal with their impact in the most appropriate manner.

Every culture has its own formal and informal means of bringing men and women together; these practices range from permitting a young man to escort a young woman to church to allowing or even encouraging the couple to live together openly. Apart from activities related to or leading to marriage, societies provide sexual outlets by legitimate or semilegitimate means such as various forms of prostitution. While society facilitates sexual activity on one hand, it also checks sexual behavior, particularly those forms considered aberrant or deviant. But what is "deviant"?

Criteria: Statistical, Medical, Legal, and Moral

Sexual behavior is usually judged by one of four criteria: statistical, medical, legal, or moral. Because these factors are often combined and interrelated, a certain practice may be illegal because it is unhealthy or rare because it is illegal. There is a tendency to bolster judgments by linking one criterion with others, sometimes quite arbitrarily. Often a past association exists despite new evidence; for example, if a certain sexual behavior has been legally prohibited because it was once considered harmful, the law may persist years later, after the medical profession has changed its mind. The use of some criteria is inevitable since human behavior must be regulated by society. But there are serious dangers in their misuse, especially if they become too interdependent or rigidly associated.

Perhaps the most easily abused of the four criteria is the statistical one. Because rare behaviors frighten us, individuals exhibiting them become objects of unusual attention. Society responds sometimes by canonizing them but more often by persecuting them. Probably it is the unpredictability of the deviant's behavior that is frightening to the rest of us. Such is the force behind this attitude that society will often delude itself into believing that a certain deviant behavior is rare so that we can deal with it more effectively. Studies that report high rates of homosexuality, for example, run into tremendous resistance. Although it is possible that a given study is erroneous in its findings or conclusions, the objection that homosexuality *could* not be that prevalent often means that it *should* not be that prevalent.

The statistical norm is misused in the other direction also by making any behavior acceptable if enough people engage in it, despite the existence of valid medical or moral objections. (The smoking of cigarettes is an example.) Only recently premarital sex was condemned on college campuses and deviants in this regard were penalized if caught. Currently peer pressure is from the other direction, and some students are made to feel "deviant" if they wish to abstain. Although moralists usually cheer when social pressure works in favor of their choices, their position is weakened in the long run when morality becomes simply a matter of prevalent custom.

The health criterion also has a sorry history as a regulator of sexual behavior. Although the medical profession has neglected the scientific study and teaching of sex, it has not refrained from speculation and ad hoc pronouncements, which have often influenced the shaping of societal views. Perhaps the most dramatic example was the fostering of misconceptions about masturbation. Typically, the physician is not out to do harm—he simply does not know how to do good when confronted with sexual issues at the clinical or societal level. This is now changing, fortunately, and it is only a matter of time until physicians will be as competent in handling sexual complaints as they are in other fields.

Today few informed Americans regard our sex laws as a valid criterion of sexual behavior. If our sex laws were enforced, most of us apparently would be in jail! Obviously we need laws to protect minors and to prohibit sexual acts that involve the use of force; but there is a vast difference between such necessary constraints and the antiquated and unenforceable rules purporting to regulate sexual behavior even between consenting spouses (such as which orifice of the wife's body may or may not be penetrated by the husband). For those who wish to pursue the subject of sex laws, I would recommend Donald Lunde's chapter "Sex and the Law," in *Fundamentals of Human Sexuality*; Robert Veit Sherwin's chapter "Laws on Sex Crimes," in *The Encyclopedia of Sexual Behavior*; and a collection of readings edited by Richard A. Wasserstrom entitled *Morality and the Law* (Belmont: Wadsworth Publishing Company, Inc., 1971).

One aspect of sex law that is of ongoing and active public interest involves the definition and regulation of "pornography." Although major changes have occurred, public opinion exercises enough control so that the impact of these changes varies greatly from one part of the country to another. The problem is a difficult one involving freedom of speech on one hand and the vulgar public exploitation of sex or the potential threat of more serious harm on the other.

What is the impact of reading or viewing pornographic material (whatever that may mean)? The answers fall into three categories. The first is that pornography is bad: it leads to immoral acts and sex crimes. The second finds no demonstrable effect on behavior. The third claims that pornography is useful in that it acts as a safety valve and prevents antisocial behavior by substituting fantasy for action.

Reliable evidence is scarce. The most extensive experience comes from Denmark, where all literature (but not graphic materials or film) became exempt from censorship in 1967. The result is difficult to evaluate. There has been an apparent decrease in crime rates since then, but because crime is a complex problem, it is hazardous to link the statistics to any single factor. It is significant, however, that there has been no *increase* in crime in the absence of censorship.

That America is likely to prolong its ignorance in this matter is suggested by the recent experience of the Commission on Obscenity and Pornography, which was appointed by Congress in 1967 and submitted its report in 1970 (*The Report of the Commission on Obscenity and Pornography.* New York: Bantam Books, 1970). Like all Presidential commissions, it consisted of distinguished citizens and professionals. After suffering through considerable dissension and controversy, the Commission came forth with the sweeping recommendation that "federal, state, and local legislation prohibiting the sale, exhibition, and distribution of sexual materials to consenting adults should be repealed." Publicly rejecting the report, the President called its conclusions "morally bankrupt" and chastised the Commission for performing a "disservice." The report may have been flawed, its recommendations questionable; but it is nevertheless noteworthy that in response to it there was no thoughtful discussion, no factual rebuttal, merely a categorical rejection and an emotional plea to keep America "a good country."

Sex remains a widely used yardstick of individual and communal virtue. In rejecting the report of the obscenity panel, the President was probably expressing widespread popular sentiment when he cautioned that pornography would "corrupt a society and a civilization" and that a permissive attitude might "contribute to an atmosphere condoning anarchy in every other field—and would increase the threat to our social order as well as to our moral principles."

This notion that sexual laxity will corrode and debase a nation cuts across national and political ideologies. Leaders of the Soviet Union would undoubtedly agree with the President of the United States, and official attitudes in the People's Republic of China are said to be even more stringent. Some Americans seriously believe that efforts to institute sex education in public schools are the result of "a communist plot." One wonders whether political dissidents tend to become sexually liberal or whether it works the other way round.

Did ancient Rome decline as a result of debauchery? Should modern nations therefore beware lest the same fate overtake us? I will let more learned heads worry about ancient Rome; but in my admittedly untutored historical judgment, this sexual theory of the decline and fall of nations seems to have all the profundity of a Cecil B. de Mille movie.

While the precise social impact of the public expression of sexual themes remains to be convincingly demonstrated, one could argue that the glaring discrepancies in our professed sexual ethics and actual practice are probably far more corrosive to our social fabric. We condemn "smut" and the "exploitation" of sex—yet with genteel elegance or with tasteless vulgarity we use sex to advertise and to entertain. While youngsters are exposed to all this, efforts to teach them about sex in school in a rational manner run into serious resistance. Contraceptives are not given to minors unless they are "emancipated"—for instance, by having become pregnant. We think our sex laws are ridiculous, but we continue to encarcerate some for doing what countless others do every day.

We have seen that the statistical criterion is subject to manipulation; the medical criterion has been contaminated by misconception; the legal criterion has been branded as obsolete. Then what is the status of the fourth criterion, morality? Beneath the variations in sexual standards, the very basis upon which these standards rest is a cause of disagreement. A fundamentalist Protestant may base a moral injunction on some specific Biblical passage; a devout Catholic may be guided by the teaching of the Church. In both cases the basis of the moral belief is revealed truth, and rational arguments may or may not enter into it. Less dogmatic believers or nonbelievers may be guided by philosophical or social utilitarian arguments, while still others "play it by ear" or do what they like and can get away with.

Historically the long-standing Judeo-Christian tradition has shaped Western sexual morality. The Catholic Church and conservative Protestant and Jewish groups still adhere to the traditional definitions, which restrict sex to heterosexual intercourse between spouses. The absolutist position clearly predates the Puritans and the Victorians, who are often held up as the epitome of sexual repressiveness. Although the words "Puritan" and "Victorian" are often used interchangeably in discussions of sexual morality, a close examination reveals important differences between these periods. Sex within marriage was considered quite appropriate by the Puritans, whose restrictions and harsh punishments pertained to behavior that endangered the integrity of the family. The Victorians by contrast seem to have fallen prey to the notion that there was something unhealthy about sex—a revival of earlier beliefs in the precious nature of semen and the deleterious effects of its loss.

The monogamous ideal is precise; it leaves nothing to chance or folly. Its primary shortcoming is that not many people seem to be able to abide by it. For better or for worse, marital coitus accounts for only part of the total sexual outlet of people examined in large samples. Because this absolutist stance has failed to provide adequately for need, concessions of varying orders of magnitude have become necessary. But once the lid is off, confusion has accompanied freedom with the result that alternatives to the monogamous ideal have ranged from the thoughtful exercise of relative sexual freedom to no-holds-barred, do-it-yourself morality.

The uncertainties and doubts that perplex many thoughtful men and women have pervaded the ranks of the clergy. While some clergymen cling to the traditional dictates, others shock their parishioners by condoning behaviors that conflict with the conventional standards. A serious rethinking of traditional norms is accompanied, hopefully, by attempts to retain the essence of the past while adapting it to the needs of the present.

Of contemporary writers in the field of sexual morality, the most widely known is Joseph Fletcher whose term, *situation ethics*, has become equated with the relativistic approach. As conceived by Fletcher, the morality of a sexual act is not determined *a priori* but rather in the context of the motivations of the participants and the circumstances of the event. The condition that exonerates the sexual act is agape, the highest and most unselfish form of sexual love. This ethic (or some popularized version of it—"It's okay if you're in love") is probably the prevalent sexual code among today's young adults.

The shortcomings of situation ethics are evident to thoughtful people including Fletcher himself. How do you recognize agape? How do you anticipate the future consequences of an act? Is it possible to be certain that a third party will not be hurt unwittingly? How do you avoid potential abuse, in which mere liking or lust replaces agape?

Sex and Youth

Scratch the surface of sexual conventions, moral and legal, and it becomes apparent that the central concern of society is with its youth. Moral censure and legal penalties pertaining to sex offenses against children are very severe. Child molesters are so abhorred that they are not even safe in jail but must be kept isolated because other inmates will assault them. Other prohibitions are meant to act as indirect safeguards by protecting the integrity of the family. Attempts are made to justify harsh measures against homosexuals on the grounds that they corrupt and abuse youthful innocence, a charge that has not been documented adequately.

No society can trifle with the welfare of its young and expect to survive; but protective concern for youth can be perverted to serve extraneous ends and used as a cover to perpetrate ignorance and suffering in the name of decency and morality. Statistics on venereal disease and illegitimacy indicate that our efforts to protect the young have been ineffectual.

A recent *Newsweek* article (January 24, 1972) predicted that if venereal disease continues to climb at its present rate, one in five of the children in Los Angeles will have contracted gonorrhea or syphilis by the time they graduate from high school. In the current epidemic, one in five persons with gonorrhea is under age twenty. In 1971 there were more than 5,000 cases among teen-agers between ten and fourteen and

2,000 cases among youngsters below the age of nine. At this rate it is predicted that 50 percent of the population will contract V.D. by the age of twenty-five.

The ratio of illegitimate to legitimate births has been climbing steadily during the last several years. In 1970 there were 626 illegitimate births in California to mothers under fifteen, 8,917 to mothers fifteen to seventeen, and 10,021 to mothers eighteen to nineteen. The overall total in California alone would probably approximate 20,000 births a year.

It is hardly necessary to point out that V.D. and illegitimacy are not what insurance companies would call "acts of God." Since we have effective cures for V.D. and reliable means of contraception, the fault is not in our medical technology or lack of scientific knowledge. Then what is the problem? "Promiscuity and permissiveness," some would say. "Ignorance," others would retort. Without attempting to resolve this dilemma, I would like to raise two issues that seem closely related to the problem: our notion of pre-adult sexuality and the need for sex education.

Although Freud and Kinsey did much to enhance public understanding of sex before adulthood, our basic collective view still considers the child as a sexual innocent and adolescence as the period of sexual awakening. Under this premise it is easy to see why some would fear and oppose sex education of children and youth: for children such enlightenment would be unnecessary, and for adolescents it would carry the risk of precocious stimulation leading to excessive preoccupation and actual experimentation with sex.

Although all of the findings of Kinsey and the conclusions of Freud are not universally accepted in scientific circles, for all practical purposes professionals in pertinent fields agree that sexuality starts not with puberty but with birth. From this vantage point one reaches quite different conclusions about the necessity and timing of sex education.

Earlier contributions are complemented by new data emerging from the work of Harry F. Harlow and his associates at the Wisconsin Regional Primate Research Center. Working with monkeys, Harlow has developed a model of psychosexual development that is potentially applicable to humans. Harlow views sex in the context of a "heterosexual affectional system" that develops in primates through three discrete subsystems: mechanical, hormonal, and romantic.

The subsystem of "mechanical" sex is dependent on the network of reflexes that mediate erection, orgasm, pelvic thrusting, and other responses to sexual stimulation, including certain sequences of gender-specific postures that culminate in adult copulatory patterns. The hormonal subsystem controls the appropriate development of sex organs, secondary sexual characteristics, fertility, and, in nonhuman primate females, the periods of sexual receptivity. The romantic subsystem is based on an affectional bond betwen heterosexual partners and grows out of preceding affectional interactions.

In this model heterosexual love and sex are seen as the culmination of affectional systems involving the love of the mother for the infant, the love of the infant for the mother, peer love, and paternal love. Although all these affectional subsystems are closely integrated and each is important in its own right, Harlow places particular emphasis on peer love as expressed and developed through peer play.

One of the most fascinating aspects of this research has been the outcome of social deprivation studies. Although the physical development of monkeys raised in isolation is quite normal, the affect on sociosexual behavior is disastrous. When they are sexually aroused and in the presence of receptive females, socially deprived males do not know what to do; they mount sideways, brutally assault the female, or act in some other inappropriate fashion. Socially deprived females do not know how to "present" sexually; they flee or attack the male. The conclusion is clear: hormones and reflexes notwithstanding, monkeys need other monkeys for appropriate psychosexual development.

Admittedly monkeys are not people (an argument that some defensively overuse), yet Harlow's work adds impressive evidence that sexuality is not a function that will "automatically" shift into gear during puberty or on the wedding night. Instead sex develops along with the individual and in successive phases until adult sexuality emerges in full form.

Erik H. Erikson is best known for his contributions to the concept of identity, which involves much more than sex but has a very important sexual component. While the individual gradually evolves a psychosexual identity (Am I a man or a woman? What kind of man or woman?), he uses sex as a vehicle in the quest and clarification of his overall identity. Much overtly sexual activity thus serves larger goals. Parents and society have to contend with some infuriating behaviors that result from this versatility of adolescent sexuality.

Erikson starts from the basic psychoanalytic concepts. He does not belabor sex in his writings, but his views are quite consistent with those that have been discussed above. Development, in Erikson's terms, follows an epigenetic model progressing from the simpler to the more complex with the precursors of various phases developing concurrently but at different paces.

In some societies adolescent sexual behavior seems to follow naturally from the sexual activities of children to the development of adult patterns. The view of Western society is consistent with Freud's formulation of a "latency period," a sort of sexual calm before the adolescent storm. Although the Kinsey data did not support this in the exact terms of Freudian chronology, they showed a relative decrease in total sexual outlet for both sexes before the onset of puberty.

Adolescent sexuality is many things. In psychoanalytic terms it is a final (hopefully) rehash of infantile sexuality with its many "pregenital" features. It is a period of stress but a unique opportunity for the resolution of old conflicts before the adult character structure begins to rigidify. Confronted with these resurgent impulses, the adolescent may resort to a variety of maneuvers; he may wholeheartedly yield to them or resolutely deny them. Some adolescents generalize these defenses so as to deny any pleasurable feeling, sexual or otherwise: they become ascetic. Others attempt to relieve their own inner turmoil by becoming insufferably judgmental about the behavior of others, particularly their parents.

Sex in adolescence, as in adulthood, is used for a variety of ends. It can express love; it can be an effective means to obtain love or to indicate status, popularity, membership in a group, or loyalty to a cause. Like adults, adolescents may engage in sex not for any profound psychological motive but merely because it is fun. Sex is also a potent vehicle of aggression or hate. Because adolescents are generally more vulnerable than adults, when adolescent sexuality takes a bad turn it becomes quite difficult to deal with. Sex is an effective means for adolescents to keep parents awake at night. While parents sometimes have good reason for concern, and adolescents are not above cutting off their own noses to spite their faces, one also encounters a great deal of unnecessary handwringing about the wild ways of youth. Adults often forget their own adolescent foibles or overreact because they themselves missed out on what are today recognized as acceptable degrees of sexual freedom. Parents may also be taken in by adolescents who talk big and pretend to act in ways that would support their idle talk.

It is easy for adults to draw erroneous conclusions by projecting themselves into adolescent life situations. An adult man may feel that if he went camping overnight with a woman, he would surely sleep with her; hence he assumes that intercourse is taking place whenever an adolescent couple goes camping together—which may or may not be true. Usually adults are oblivious to the anxieties, inner doubts, and turmoils that sexually paralyze youngsters. A pubescent girl would sometimes rather die than show herself naked to her boyfriend, not necessarily because of moral considerations but because she is self-conscious about her sprouting pubic hair or asymmetrical breasts. Similar concerns affect boys. Although it is not possible to say whether such considerations constitute the exception or the rule today, it is reasonable to assume that whatever changes in sexual mores are taking place, adolescents are still adolescents and still subject to these concerns.

Sex Education

There are many areas in which concerns about sex, youth, and morality interact with the fabric of society. Such interaction becomes particularly intense and conflicted around the issue of sex education, a subject that appears erratically in the curriculums of our educational institutions. This is true for elementary schools, high schools, colleges and universities, and even for medical schools. Here and there a school or college offers a serious course or sex education program, but by and large the topic is missing or buried and camouflaged. I have never heard of a degree program in human sexuality, an endowed chair, or even a simple professorship.

This anomaly in scholarship and teaching surely cannot be explained by any rational considerations, because there are few areas of life that touch everyone so profoundly, and interest in sex is hardly lacking. This preoccupation is well known and relentlessly exploited, yet no legitimate avenues are made available for the gratification of this curiosity in an intelligent manner. Sex has simply not been considered a fit subject for academic pursuit.

The roots of this attitude are not entirely clear to me. It is too widespread to be explained by "Victorianism" or other cultural scapegoating. The clash of proponents and opponents of sex education is most dramatic when it involves the instruction of children and adolescents. In colleges and universities the lack of academic interest is less often due to the opposition of external forces and more often the result of inertia, the lack of readily available teaching materials, and fear of interdisciplinary trespassing. Since sex is recognized to have biological, psychological, and cultural components, specialists in any one area are reluctant to take on the topic as a whole, and interdisciplinary teaching is more easily advocated than practiced.

Some epic battles have been fought over sex education at the elementary and high school levels. One of the more celebrated took place in Anaheim, California, and has been instructively chronicled by Mary Breasted in *Oh! Sex Education!* (New York: The New American Library, 1971). This book, which received favorable reviews, reveals fascinating aspects of the people who become embroiled in these conflicts.

Opposition to sex education may arise from thoughtful concern about the manner in which the topic is to be presented as well as from the fear of some fantastic conspiratorial scheme underlying such programs. But more often the nagging questions are: Will sex education precociously stimulate the youngsters? Will it lead to premarital or deviant experimentation? There is no conclusive evidence to date, and some of the people who decry this lack are in fact responsible for making it difficult to gather such evidence.

The conservative attitude is not necessarily the safer one. Surgeons who do not intervene because of the operative risks involved sometimes end up with dead patients. Just as the risks of using contraceptives must be weighed against the risks of not using them, the potential hazards of sex education must be evaluated with the dangers of sexual ignorance. We cannot blame sex education for the soaring rate of illegitimate births or the spread of venereal disease, because sex education does not currently exist as a major influence.

One should not by the same token oversell the potential value of sex education as a deterrent to venereal disease, premarital sex, or any other real or fancied problem. There are many reasons why a teen-ager becomes pregnant; ignorance of contraception is one reason, and sex education will dispel ignorance. It may even help the youngster to understand some of the psychological forces involved in sex. But some girls are still going to get pregnant, because adolescents, like adults, are not always motivated by exclusively rational considerations.

Incidentally, controversial or threatening subjects seem to bring out the "scientific" best in people. I have never heard anyone demand documentation that the teaching of American history makes students patriotic or produces responsible citizens. But if we are going to teach students about menstruation, which affects at least half of mankind, or about various forms of sexual behavior, which affect all of mankind, we must document our aims with precision, proof, and certainty.

Opposition to sex education is rarely absolute; some recognition is given to the need for instruction, while controversy centers upon the questions of where, when, and by whom. Those who oppose the schools' doing the teaching propose that parents should enlighten their children in a context and manner consistent with their style of life and ethical outlook. The drawback is that few parents are eager or competent to take on this task. Other problems pertain to the difficulty of planning and implementing a sex education program. The issue has been neglected for so long that the prospects of introducing sex into a modern curriculum seem akin to trying to integrate a primitive piece of machinery into a technically sophisticated plant.

The way out of this dilemma will start with the recognition that, while there is much that we do not know, nothing prevents us from learning about sex just as we learn about other physiological or emotional phenomena. While we stand paralyzed or proceed with excruciating caution, children and adolescents are learning about sex in the traditional haphazard manner. The question is, can we influence this on-the-job learning?

We need useful, reliable information about sex and gifted people to gather such data. With some notable exceptions, the cream of the nation's minds is not found in this field. While it is true that sex research has become quite lucrative, and hardly anyone now faces public ridicule or legal censure, it is still less than fully respectable. As long as this is true, gifted scientists, who are subject to the same reward systems as other people, will stay out of the field, and our information will remain spotty and incomplete.

Sex in 2001

Novelists who dwell in the brave new world of tomorrow often portray the gratification of sexual desire by efficient mechanical or chemical means without emotional encumbrance. Barbarella, the science fiction superastronaut female, simply takes a pill to experience a superorgasm. The increasing availability and use of mechanical vibrators could be interpreted as evidence that today's satire might become tomorrow's reality. We are still far from the day when our bedside clock radios and electric blankets may be supplemented by gadgets providing erotic convulsions on demand, timed and modulated to fit the needs of the moment; but the progressive mechanization of sex is of genuine concern to authors like Rollo May, who recognize a certain danger of depersonalization inherent in the Masters and Johnson approach.

Any trend or movement that looks like a departure from age-old custom may turn out to be the recycling of the same old practices. The hippie, who was hailed as the New Man a few years ago, has been compared to his Bohemian antecedents by Bennet M. Berger under the title, "Hippie Morality—More Old Than New." (In J.H. Gagnon and W. Simon, eds., *The Sexual Scene.* Chicago: Aldine Publishing Company, 1970.)

It is difficult—if not impossible—to predict what new technological inventions or drastic social development may change human behavior substantially. It is less dramatic but more plausible to project existing trends. Remember that the cumulative effect of progressive change may become startling at a certain threshold.

It would seem reasonable to assume that current trends toward sexual permissiveness will continue during the foreseeable future. Compared to Kinsey's data, statistics gathered in a survey by Vance Packard (*The Sexual Wilderness*. New York: Pocket Books, 1970) indicated a gradual increase in the number of college girls who have had premarital intercourse: at age twenty-one, for example, the ratio had risen by almost 60 percent. In Kinsey's sample about 27 percent of college-educated females had had premarital coitus. Packard revised this figure upward to 43 percent. This is compared to 63 percent of English university women, 60 percent of Germans, 54 percent of Norwegians, 35 percent of Canadians, and 10 percent of Italians. Other surveys (at Oberlin, for example) have generally supported Packard's figures. As a general estimate, about half of the unmarried women presently in college will probably engage in coitus before graduation. Of course, college girls constitute only part of the population and by no means a representative one.

Male behavior seems to have changed less—mostly because there was not as much room left for change. In the Kinsey male sample, premarital sex was reported by 98 percent of elementary school graduates, 85 percent of high school graduates, and 68 percent of the college group.

How far will these trends continue? Will premarital virginity soon become another historical relic? More than three decades ago L.M. Terman predicted that, if the progressive decline in premarital virginity continued, few males born after 1930 and few females born after 1940 would be virgins when married. This expectation does not seem to have been fulfilled as yet. Ira L. Reiss points out in "How and Why America's Sex Sandards Are Changing" (in J.H. Gagnon and W. Simon, eds., *op. cit.*) that the more important changes in premarital sexual patterns have been in attitudes rather than in behavior. Despite more permissive attitudes, some young people will hold out for religious, moral, or psychological reasons and a variety of other motives. These persons may find themselves under considerable peer pressure. A recent booklet, *Sex and the Yale Student* by the Student Committee on Human Sexuality, already finds it necessary to reassure the reader that there is nothing necessarily wrong with a person if he or she does *not* feel ready to have intercourse while in college.

There is no evidence that college students have become indiscriminately promiscuous. The Yale booklet reports there is some active experimentation at the beginning of the school year, but most students soon settle down to more stable relationships. Only a minority continues the pattern of multiple involvements. There is reason to believe that most couples will continue to expect an emotional tie before they undress for each other.

The key issue regarding premarital sex (as well as extramarital sex) hinges on the definition of marriage and the family. There was a time when only the church could legitimize a marital union, but now we have civil marriages. What next? Proposals and predictions about the future of marital relations include trial marriages, renewable marriage contracts, more unabashed serial monogamy, mutual polygamy, single-parent families (in which single men or women rear their own children), unstructured cohabitation (better known as "shacking up"), communal living, and polygamy for senior citizens to solve the plight of the more numerous elderly women.

Today a large number of marriages end in divorce and the apparent deterioration in the stability of the American family is one factor reflected by increasing interest in alternative life-styles. Isolated experiments to abolish marriage in various utopian communities have been generally short-lived. Although the perfectionist Oneida Community in upstate New York lasted for about thirty years in the middle of the nineteenth century, most group marriages do not remain viable for an extended period of time. Such experiments, which have been conducted and abandoned on a larger scale in the Soviet Union and elsewhere, have their modern counterparts in communes and other novel living groups. Unmarried young people are certainly living together more openly today. Are these arrangements negations of the concept of marriage or new forms of marriage based on more equalitarian principles, revised concepts of sex roles, and alternative forms of child-rearing?

This view toward 2001 has so far been centered upon heterosexual intercourse. Chances are, we can expect a gradual liberalization of society's attitudes toward other forms of sex. As indicated earlier, there is already a more open and unabashed endorsement of auto-erotic activity as a supplemental source of orgasm, and homosexuality is becoming more overt. Although the threat of increased or more blatant homosexuality is another of this nation's nightmares, the laws pertaining to acts between consenting adults are likely to become liberalized in this country as they now are in England. It is highly unlikely that homosexuals are about to take over control of society or even a substantial segment of it, entrenched as they may be in some enclaves.

Predictions about the future of sex occasionally surface in magazines. In a rather uninformative article in *Look* (July 25, 1967), Marshall McLuhan and George B. Leonard opined, "Sex As We Know It

May Soon Be Dead." Why? Because males and females are beginning to look and act alike. This notion of "unisex" is primarily a concept of the fashion industry and as likely to affect the man and woman in the street as are the advertisements in *Vogue*. Occasionally someone comes up with a nightmare such as Stanley Kubrick's *Clockwork Orange* in which sex and aggression become intermixed in a streamlined world of advanced technology.

Although permissiveness expresses the dominant theme in this guessing game, other voices raised in counterpoint assert the belief (and perhaps the hope) that the tide will turn. Witness the remarkable religious revival among the young that denounces the ways of the flesh. Is this another fad or a true indicator of the future?

There is a growing realization that we may be coming around full circle, that endless vistas of flesh ultimately produce a soporific effect. In *Fundamentals of Human Sexuality*, Eitner quotes Diderot, the eighteenth century critic, who reviewed annual salons to the point of satiety. Diderot wrote, "I am no Capuchin, but I confess that I should gladly sacrifice the pleasure of seeing any more attractive nudities, if I could hasten the moment when painting and sculpture having become more decent will compete with the other arts to inspire virtue and purify manners. I think I have seen enough tits and behinds. These seductive things interfere with the soul's emotions by troubling the senses."

Apart from surfeit, there is a widening concern that "sexual freedom" does not come free. In a sensitive *New Yorker* essay, "Permissiveness and Rectitude" (February 28, 1970), Leopold Tyrmand wrote of the "treasure of shame" and the cost of our loss of innocence.

If hippie morality was "more old than new," if nudity waxed and waned in the eighteenth century French salon, if group marriage was tried and abandoned in nineteenth century New York, if permissive and restrictive attitudes can be represented by the swing of the pendulum, one might conclude that nothing is really different today except the establishment of sex research and the beginnings of sex education, which hold forth the promise that future men and women will be less ignorant of their sexual function.

But one other factor is truly unique in our time—and compared to this factor, issues such as censorship or tolerance of public nudity are trivial. Without this factor it would be impossible to reevaluate premarital sex, trial marriage, or extramarital relations, or to project the many portentious predictions that fill some hearts with hope and others with horror. This single most significant factor is the availability of reasonably safe and effective contraceptives to wide segments of the population. For the first time in human history, sex and reproduction can be readily separated. The separation of these functions will have profound repercussions upon our sexual conventions, morals, and laws.

Carl Djerassi of Stanford, who did the basic chemical research that led to the birth control pill, does not foresee new major breakthroughs in contraception in the immediate future, but expects improvement and refinement of current techniques and materials. Nonetheless, wider use of current contraceptives, along with changes in attitudes and laws on abortion, will exert an influence that begins with the status of women and the redefinition of gender roles, then extends far beyond sexual matters to the economic and political foundations of society. So vast are the implications that our ability to control reproduction has been compared to the discovery of fire or nuclear energy in its significance for mankind.

READER'S GUIDE

CHAPTER ONE: The Physical Basis of Sex

Anatomy of Sex Organs

There are innumerable books on anatomy at all levels of complexity. For the next level of detail I would suggest: Dienhart, C.M. *Basic Human Anatomy and Physiology*. Philadelphia: W.B. Saunders Company, 1967. The anatomy chapter in *Fundamentals of Human Sexuality* is another worthwhile reference.

More specialized works: Netter, F.M. *Reproductive System*. The Ciba Collection of Medical Illustrations. Vol. 2. Summit, N.J.: Ciba, 1965. Netter's multicolor plates are famous; the text is condensed. Dickinson, R.L. *Atlas of Human Sex Anatomy*, 2nd ed. Baltimore: The Williams and Wilkins Company, 1949, has detailed drawings in black and white. It is not as polished as Netter's book but contains much information on variations in sizes and shapes of sexual organs. Many detailed anatomy textbooks are available. (As a medical student, I used *Gray's Anatomy*, but I can hardly claim a sentimental attachment to what I remember as an ordeal by fire.)

Sex Hormones

College biology texts often have interesting material on genetics and hormones. A good general introduction to the functions of the body including the reproductive system is: Guyton, A.C. *Function of the Human Body*. 3rd ed. Philadelphia: W.B. Saunders Company, 1969. Both *Fundamentals of Human Sexuality* and *The Encyclopedia of Sexual Behavior* have specific chapters on sex hormones. There is a lucid discussion also in Crawley, L.Q., J.L. Malfetti, E.I. Stewart, and N. Vas Dias. *Reproduction, Sex and Preparation for Marriage*. New Jersey:

Prentice-Hall, 1964. I would also recommend Harrison, R.J. *Reproduction and Man.* New York: W.W. Norton and Company, 1967.

More detailed and sophisticated accounts are found in general endocrinology texts: Williams, R.H. *Textbook of Endocrinology.* Philadelphia: W.B. Saunders Company, 1968; Turner, C.D., and J. T. Bagnara. *General Endocrinology.* 5th ed. Philadelphia: W.B. Saunders Company, 1971. There are also several books that have a wide collection of essays on various aspects of sexual endocrinology and behavior, and many of these essays are intelligible to educated laymen: Money, J. (ed.), *Sex Research: New Developments.* New York: Holt, Rinehart, and Winston, 1965; Beach, F.A. (ed.), *Sex and Behavior.* New York: John Wiley and Sons, 1965; Young, W.C. (ed.), *Sex and Internal Secretions.* 2 vols. Baltimore: The Williams and Wilkins Company, 1961; Diamond, M. (ed.), *Perspectives in Reproduction and Sexual Behavior.* Bloomington: Indiana University Press, 1968.

Sexual Response

This is mostly Masters and Johnson territory. There is little material about specifically sexual matters, such as orgasm, even in medical physiology textbooks. There are, however, some informative (though somewhat dated) chapters in Kinsey's *Sexual Behavior in the Female.*

The original Masters and Johnson report has reached a wide readership despite its awkward prose and has been quoted and summarized extensively: Masters, W.H., and V.E. Johnson. *Human Sexual Response.* Boston: Little, Brown and Company, 1966. A summary that is adequate for most purposes is: Brecher, R., and E. Brecher. *An Analysis of Human Sexual Response.* New York: New American Library, 1966. This paperback includes some additional articles of interest. *Fundamentals of Human Sexuality* has a chapter on the physiology of sexual functions.

Readings from *Scientific American* on the biological basis of behavior have been collected in: *Psychobiology.* San Francisco: W.H. Freeman and Company, 1967. Especially pertinent are two articles in this collection: Levine, S. "Sex Differences in the Brain" (pp. 76-81) and Olds, J. "Pleasure Centers in the Brain" (pp. 183-88). These articles can be obtained individually from the publisher.

Finally, Julian M. Davidson and Gordon Bermont have a book under preparation provisionally entitled *Biological Bases of Sexual Behavior* (Harper and Row Publishers, Inc.) which will undoubtedly live up to the promise of its title. It will be published as a paperback. I certainly look forward to its appearance.

CHAPTER TWO: Sexual Behavior

Varieties of Sexual Experience

The statistics of sexual behavior, like all statistics, tend to appear forbidding, but if one approaches these figures a few at a time with some understanding of how to interpret them, they can be quite fascinating. The Kinsey data are the chief source of the sex statistics one encounters in the literature, even though they are not always so credited.

Chapter 7 of *Fundamentals of Human Sexuality* deals with the frequency and components of sexual outlet. If you are seriously interested in these topics, you must of course go to the original Kinsey volumes: Chapter 6 of *Sexual Behavior in the Male* and Chapter 13 of *Sexual Behavior in the Female.*

The better descriptions of the lives and loves of sexually enterprising people are in the general literature rather than in the scientific and hopefully scientific works on sex. Strother B. Purdy has written, in my opinion, an excellent introduction to the erotic in literature (Chapter 14 in *Fundamentals of Human Sexuality*). It includes a carefully selected and annotated bibliography. A book examining a few select erotic works in greater depth was favorably received by critical reviewers: Marcus, Steven. *The Other Victorians.* New York: Basic Books, 1966. (A Bantam paperback edition was published in 1967.)

I have found *My Secret Life*, by an anonymous Victorian gentleman, the most instructive biography of a man relentlessly driven by sex. For most readers the abridged paperback edition (New York: Grove Press, Inc., 1966) is long enough—696 pages. Please remember that practically all of this literature has been written by males predominantly for male consumption.

Enhancement of Sexual Pleasure

The best-known ancient love manuals were translated by Sir Richard Burton in the late nineteenth century but have become widely available only during the past decade or two. These books make fascinating reading but can also be puzzling and tedious. They combine perceptive insights into human sexual nature with utter nonsense. I find *The Perfumed Garden* by Shaykh Nefzawi the most interesting, although the *Kama Sutra* by Vatsyayana is more famous. Both can be obtained in paperback (New York: G.P. Putnam's Sons).

Japanese "pillow books" were erotic manuals kept beneath the pillows of bridal beds, and until the 1930s some Japanese department stores placed copies of these in bridal chests. These books are not generally available, but excellent examples of such art—the *shunga* or "spring pictures"—can be found in: Rawson, Philip. *Erotic Art of the East.* New York: G.P. Putnam's Sons, 1968. This volume covers Indian, Chinese, and Islamic erotic art also and is profusely illustrated with art of high quality.

The prototype of the contemporary marriage manual is Theodoor Van de Velde's *Ideal Marriage*, first published in 1926. Although it has sold more than a million copies in this country alone, over the years it has become hopelessly dated. Recently revised by a British physician, it still retains its quaint flavor and some of its inaccuracies. If you feel you are likely to find the more explicit manuals offensive, however, *Ideal Marriage* may be worth your while.

A more recent, prolific, and widely read author is Albert Ellis, a New York psychologist. Of his various works, the one that most resembles a sex manual is: *The Art and Science of Love.* New York: Lyle Stuart, Inc., 1965, paperback. In the 1950s Ellis was considered avant-garde in his outspoken endorsement of sex outside as well as inside marriage, but his writings appear sedate compared to the language of today. Ellis is a well-informed professional whose books reflect prevalent knowledge even though one may take issue with his sexual philosophy. Unlike many sex manual authors, he documents his statements and lists his sources.

The reliable and sober information that you would expect to learn from a well-informed physician is available in this book by a professor of psychiatry: Hastings, Donald W. *Sexual Expression in Marriage.* New York: Bantam Books, 1967, paperback. *Patterns of Sexual Behavior* by Ford and Beach has an informative chapter on circumstances for coitus, with cross-cultural and cross-species perspectives.

Catholics may be interested in a gynecologist's book that has a foreword by and carries the imprimatur of Richard Cardinal Cushing: Lynch, William A. *A Marriage Manual for Catholics.* New York: Pocket Books, 1968.

Since one cannot have too much of a good thing, we can probably look forward to a continuing series following *The Sensuous Woman* and *The Sensuous Man* (New York: Lyle Stuart, 1969 and 1971 respectively). There is, for example: Chartham, Robert. *The Sensuous Couple.* London: Ballantine Books, 1971. A precursor of these is: Eichenlaub, John. *The Marriage Art.* New York: Dell Publishing Company, 1961, paperback. Manuals of this type make no pretense at scholarship; there are no references, no documentation, and no credits. Even though the author is sometimes a physician, the style and approach are quite personal. Typically, these books contain a mixture of truths, possible truths, and errors. Although millions upon millions read these books, presumably to be informed, it is impossible to estimate how many actually try out the advice.

Another type of sex manual, the "picture book," consists of actual photographs of love-making, but not necessarily lurid shots or genital close-ups. If anything, the models tend to be languorously limp and obviously unaroused. A good example of this type of book, and one that has an informative text, is: Harkel, Robert L. *The Picture Book of Sexual Love*. New York: Cybertype Corporation, 1969.

If you are interested in a quick exposure to writers in this field, I suggest a paperback that includes sixty-three articles by more than thirty authors—actually brief articles reprinted from *Sexology* magazine: Rubin, Isadore (ed.), *Sexual Freedom in Marriage*. New York: Signet Books, 1969. I am reasonably certain I have not omitted any books that are head and shoulders above those listed, but I am equally sure there must be quite a few that are comparable. I must confess I am not wild over most of this literature.

Even though it is not a typical sex manual, it would be impossible to omit mention of *Everything You Always Wanted to Know About Sex (But Were Afraid to Ask)* by David Reuben (New York: David McKay Company, Inc., 1970), a book that has apparently made publishing history and is fascinating from that point of view. Its best features are probably its ingenious title and format, although the question-and-answer idea is not new. Le Mon Clark, the question-and-answer man of *Sexology* magazine, has an earlier book called *101 Intimate Sexual Problems Answered*. Reuben's book has received bad reviews for good reasons. More recent, far superior, and by far the most sophisticated book in this entire general category is *The Joy of Sex*, edited by Alex Comfort. New York: Crown Publishers, 1972. This is a literate and artfully illustrated volume.

As examples of less flamboyant books, not so preoccupied with technique but more concerned with the broader psychological issues of the sexual relationship, I would suggest: Berne, Eric. *Sex in Human Loving*. New York: Simon and Schuster, 1970; Duyckaerts, Francois. *The Sexual Bond*. New York: Delacorte Press, 1970; May, Rollo. *Love and Will*. New York: W.W. Norton and Company, Inc., 1969.

Sexual Malfunction

Because practically all the marriage and sex manuals deal to some extent with problems of impotence and frigidity, the references listed in the preceding section are pertinent here.

The work of Masters and Johnson in this area is well summarized in a paperback: Belliveau, Fred, and Lin Richter. *Understanding Human Sexual Inadequacy.* New York: Bantam Books, Inc., 1970. The primary source is: Masters, W.H., and V.E. Johnson. *Human Sexual Inadequacy.* Boston: Little, Brown and Company, 1970.

If you are curious about how psychotherapists approach sexual problems, a concise and cogent introductory work is available: Colby, Kenneth M. *A Primer for Psychotherapists.* New York: The Ronald's Press Company, 1951. Behavior modification techniques are described in: Wolpe, Joseph, and Arnold A. Lazarus. *Behavior Therapy Techniques.* New York: Pergamon Press, 1967.

The sexual life of older persons is in the process of reevaluation, but writings on this subject are sparse. A new book by Simone de Beauvoir is slated for publication by G.P. Putnam's Sons. Entitled *The Coming of Age,* it brings together some of the Masters and Johnson findings on the sexual response of older persons with various literary references to this theme.

Variations and Deviations of Sexual Behavior

The classic on sexual deviations is Richard von Krafft-Ebing's *Psychopathia Sexualis* written in 1899 and now considered of historical interest only. Among contemporary works I would recommend an interesting anthology with writings by well-known authors of various theoretical persuasions: Ruitenback, H.M. (ed.), *The Problem of Homosexuality in Modern Society.* New York: E.P. Dutton and Company, Inc., 1963. A serious and well-balanced work is: West, D.J. *Homosexuality.* Chicago: Aldine Publishing Company, 1967. Interviews with various types of homosexuals and with some of the ranking specialists in the field of sex are included in a 600-page review that is nontechnical but thoroughly researched and comprehensive: Karlen, A. *Sexuality and Homosexuality.* New York: W.W. Norton Company, Inc., 1971. A clinical source book edited by a psychoanalyst is: Marmor, Judd. (ed.), *Sexual Inversion: The Multiple Roots of Homosexuality.* New York: Basic Books, 1965. A chapter by Evelyn Hooker examines the illness versus life-style controversy, which is discussed also in: Bieber, Irving, *et al., Homosexuality.* New York: Vintage Books, 1962.

Discussions of causes and treatment are predicated upon the underlying assumption as to what homosexuality is; if one accepts it as a life-style, there are no "causes" to talk about and there is no question of treatment. A recently published study raises again the issue of a possible genetic or hormonal basis for the condition: Kolodny, R.C., W.H. Masters, J. Hendryx, and G. Toro. "Plasma Testosterone and Semen

Analysis in Male Homosexuals," *New England Journal of Medicine* (November 18, 1971), 1170-74.

Newspaper articles are probably the best source for current developments in the homosexual world. In a city like San Francisco, the doings of the gay crowd make front-page news: "Cops Meet the Gays," *San Francisco Chronicle*, May 10, 1971, p. 1; "The Gay Bars," *op. cit.*, December 6, 1971, p. 1. Ultimately there is no substitute for an intelligent discussion with a homosexual, but that is not always easy. Since the more outspoken tend to be chauvinistic, you will soon find yourself listening to the refrain, "Julius Caesar was a homosexual, Michelangelo was a homosexual, (etc.) . . ."

In *Fundamentals of Human Sexuality*, Chapter 11 provides a general introduction to variant and deviant patterns of sexual behavior; Chapter 14, a discussion of the deviance theme in the general literature; and Chapter 16, a survey of legal aspects.

Indispensable for the reader with special interest is: Gebhard, P.H., J.H. Gagnon, W.B. Pomeroy, and C.V. Christenson. *Sex Offenders*. New York: Harper and Row, 1965. Almost any general text in psychoanalysis or psychiatry will offer ample material on sexual deviations, including the treatment; for example: Fenichel, O. *The Psychoanalytic Theory of Neurosis*. New York: W.W. Norton and Company, 1945; Redlich, F.C., and D.X. Freedman. *The Theory and Practice of Psychiatry*. New York: Basic Books, 1966. Again the cross-cultural and cross-species perspective is provided by Ford and Beach, *Patterns of Sexual Behavior*, Chapter 7.

In an entirely different class is a paperback that details the operations of sex racketeers and peddlers of "hard-to-get" items: Thompson, G. *Sex Rackets*. Cleveland: K.D.S. Corporation, 1967. Leafing through the back pages of a low-grade "girlie" magazine will accomplish the same task, a glimpse of the seedy sexual underground. Going through an entire "sex shop" requires considerably more dedication.

CHAPTER THREE: Sex and Society

Criteria: Statistical, Medical, Legal, and Moral

In Chapter 17 of *Fundamentals of Human Sexuality*, Donald Lunde surveys the historical record starting with the Talmudic and Biblical views on sex, followed by references to St. Augustine, St. Thomas Aquinas, Luther, Calvin, and more recent writers.

Joseph Fletcher's books are *Situation Ethics: The New Morality*

(1966) and *Moral Responsibility: Situation Ethics at Work* (1967). Philadelphia: Westminster Press. For a review of recent statements by various Protestant churches, see: Genné, W.H. (ed.), *A Synoptic of Recent Denominational Statements on Sexuality*. New York: National Council of Churches, available from the Department of Educational Development, 475 Riverside Drive, Room 711, New York, N.Y. 10027. The B'nai B'rith Hillel Foundations, Inc., presented a Jewish viewpoint in: Borowitz, E.B. *Choosing a Sex Ethic*. New York: Schocken Books, 1969. The Catholic perspective is presented in Lynch's *A Marriage Manual for Catholics*.

Philosophical works recommended are: Dewey, John. *Theory of the Moral Life*. New York: Holt, Rinehart, and Winston, Inc., 1960; Thielicke, H. *The Ethics of Sex*. New York: Harper and Row, 1964.

Sex and Youth

Two kinds of material are relevant to sex and youth: books to guide parents and books for the youngsters themselves to read. But with the ready availability of all manner of books and the compulsion to read the most "adult" version, many teen-agers may prefer to read Masters and Johnson. Is that bad? Although I cannot think of any specific harm such reading would do, I would not claim that my ignorance should constitute sufficient safeguard.

If your teen-ager wants to read any current book, chances are there is nothing you can do to stop him. You have the choice of forcing the activity underground (and not being party to the "crime") or letting it take place in the open, in which case you will know what is going on and can discuss the literature with your child. I would opt for the latter alternative and make the key criterion the truthfulness and reliability of the reading matter, not how "advanced" or "suitable" it is. Youngsters determine their own levels of tolerance: if a book bothers or bores them, they will let it alone. This mechanism may not be foolproof, but it is functional.

Authors like Dr. Spock, who advised parents how to teach "the facts of life" in his famous *Baby and Child Care* (New York: Pocket Books, 1970), combine well-known psychological principles with common sense. Any of the following may be helpful to parents and children: Spock, Benjamin. *Teenager's Guide to Life and Love*. New York: Pocket Books, 1971; Ginott, H. *Between Parent and Child* (1965) and *Between Parent and Teenager* (1969). New York: Avon Books; Fraiberg, Selma F. *The Magic Years*. New York: Charles Scribner's Sons, 1959. A book recommended for children ages three to eight is: Andry, Andrew C., and Steven Schepp. *How Babies Are Made*. New York: Time-

Life Books, 1968. Additional books for teen-agers are: Pomeroy, Wardell B. *Boys and Sex* (1968) and *Girls and Sex* (1969). New York: Delacorte Press; and Bohannan, Paul. *Love, Sex and Being Human.* New York: Doubleday and Company, Inc., 1969.

Publications intended for parents, teachers, and counselors are offered by the Sex Information and Education Council of the United States. Twenty articles on various aspects of sex are included in: Broderick, C.B., and Jessie Bernard (eds.), *The Individual, Sex, and Society.* New York: SIECUS, 1969. A bibliography, *Selected Reading in Education for Sexuality*, may be obtained by sending a self-addressed envelope to SIECUS, 1855 Broadway, New York, N.Y. 10023.

Sex in 2001

"Hippie Morality—More Old Than New" by Bennet M. Berger and "How and Why America's Sex Standards Are Changing" by Ira L. Reiss may be found in: Gagnon, J.H., and W. Simon (eds.), *The Sexual Scene.* Chicago: Aldine Publishing Company, 1970. For a description of changing patterns of sexual behavior in the college population, see: Packard, Vance. *The Sexual Wilderness.* New York: Pocket Books, 1970. If you are interested in reading about group sex, a reliable report is now available: Bartell, Gilbert D. *Group Sex.* New York: Peter H. Wyden, Inc., 1971. The potential significance of the birth control pill is examined in: Montagu, Ashley. *Sex, Man and Society.* New York: Tower Publications, Inc., 1969.

INDEX

ABOUT THE AUTHOR

An honors graduate of the American University of Beirut Medical School, Dr. Herant Katchadourian first came to the United States in 1958 to begin a residency in psychiatry at the University of Rochester. Following his residency, he spent a year at the National Institute of Mental Health in Bethesda, Maryland, before returning to Lebanon with a United States Public Health Service grant to conduct studies in psychiatric epidemiology. In 1966 he joined the Department of Psychiatry at the Stanford University Medical School and was coordinator of its residency program until 1970, when he became the university's first ombudsman. Dr. Katchadourian is currently Associate Professor of Psychiatry and a University Fellow at Stanford. He is author (with D.T. Lunde) of *Fundamentals of Human Sexuality* (1972).